Decolonial Pedagogy

Njoki Nathani Wane · Kimberly L. Todd
Editors

Decolonial Pedagogy

Examining Sites of Resistance, Resurgence, and Renewal

Editors
Njoki Nathani Wane
Ontario Institute for Studies
 in Education
University of Toronto
Toronto, ON, Canada

Kimberly L. Todd
Ontario Institute for Studies
 in Education
University of Toronto
Toronto, ON, Canada

ISBN 978-3-030-01538-1   ISBN 978-3-030-01539-8   (eBook)
https://doi.org/10.1007/978-3-030-01539-8

Library of Congress Control Number: 2018956819

Cover illustration: Pattern © Melisa Hasan

This Palgrave Pivot imprint is published by the registered company Springer Nature Switzerland AG
The registered company address is: Gewerbestrasse 11, 6330 Cham, Switzerland

We dedicate this work to our collective ancestors those who are living, those who have passed on, and our chthonic ancestors. We also dedicate this book to those seeking the Awakening force, may you be guided by the love of the Earth.

Acknowledgements

We would like to acknowledge that a more extensive version of this chapter: The University as a Neoliberal and Colonizing Institute; A Spatial Case Study Analysis of the Invisible Fence between York University and the Jane and Finch Neighbourhood in the City of Toronto by Ardavan Eizadirad was initially published in the *Journal of Critical Race Inquiry* in March 2017.

This acknowledgement is primarily an acknowledgement of what has come before us and the supporters that have enabled us to flourish. Thank you to our ancestors, especially our grandparents and most especially our grandmothers who have overcome extreme adversity to enable us to flourish-their love transcends time, space, and even death. To our parents whose love, support, and resilience are unconditional, we are forever grateful to you. To our siblings who find incredible ways to demonstrate their love and support of us. To our extended families' uncles, aunts, cousins who provide us with the comfort of home wherever we go. To our friends whose laughter, generosity, and compassion sustain us. To our families that were created on the basis of our relationships and not blood, we offer you our gratitude. To our children and our future generations, you are that reasons that we do this work. To our Mother the Earth for her all-encompassing love, sustenance, and beauty, may we honour you in all that we do.

धन्यवाद/Deu borem korm/Thank you

CONTENTS

NOTES ON CONTRIBUTORS

Glenn Adams is Professor of Psychology and Director of the Cultural Psychology Research Group at the University of Kansas. His work applies decolonial perspectives of cultural psychology to topics of collective memory, knowledge, and ignorance.

Ardavan Eizadirad is an instructor at Ryerson University in the School of Early Childhood Studies in Faculty of Community Services. He is also a Ph.D. candidate in the Department of Curriculum, Teaching, and Learning at Ontario Institute for Studies in Education (OISE) at University of Toronto. His Ph.D. thesis is examining the effects of standardized testing on subjective experiences of grade 3 students and parents in the context of Toronto, Canada. He is an educator with the Toronto District School Board and a community activist with the non-profit organization Youth Association for Academics, Athletics, and Character Education (YAAACE). His research interests include standardized testing, systems of accountability, community engagement, anti-oppressive practices, critical pedagogy, social justice education, resistance, subversion, and decolonization. His recent publications include "The University as a Neoliberal and Colonizing Institute: A Spatial Case Study Analysis of the Invisible Fence between York University and the Jane and Finch Neighbourhood in the City of Toronto" (March 2017), "Comparative Analysis of Educational Systems of Accountability and Quality of Education in Ontario, Canada and Chile: Standardized Testing and its Role in Perpetuation of Educational Inequity" (August

2016), Is It "Bad" Kids or "Bad" Places? Where Is All the Violence Originating From? Youth Violence in the City of Toronto, (April 2016), and "International Experience in a Non-Western Country, Teacher Habitus, and Level of Inclusion in the Classroom" (March 2016). He is currently working on a manuscript with Dr. John Portelli about Subversion & Standardized Testing (forthcoming 2018).

Chizoba Imoka is a Ph.D. candidate at the University of Toronto's Ontario Institute for Studies in Education. Her doctoral research explores student experiences in Nigerian secondary schools and seeks to make policy/practice recommendations for decolonial education reform. Within the community, Chizoba is an acclaimed advocate for public education reform in Africa, social justice and transformative youth engagement. She is the Founder/CEO of Unveiling Africa, a non-profit that provides a platform for Nigerian youth to participate in community mobilization and political leadership.

Eucharia Kenya is the Deputy Vice-Chancellor in charge of Planning, Administration and Finance, at the University of Embu. She is also a consultant on Biotechnology, Biosafety, and Biosecurity issues with international experience and expertise. Previously she taught at the Department of Biochemistry and Biotechnology, Kenyatta University, Nairobi, for over 10 years. During this period, she supervised over 30 postgraduate students and published several peer-reviewed papers. She has attracted substantial grants that have benefited many students in the greater East African Region and fostered collaborative projects and partnerships that have contributed in strengthening capacity in different aspects of biochemistry and biotechnology.

She is a very active player in the agricultural biotechnology sector in Kenya and served as a member of the steering committee under the Ministry of Agriculture, that developed the "Agricultural Biotechnology Awareness Strategy for Kenya". She was deeply involved in the discussions that culminated in the enactment of the Kenya Biosafety Act (2009) and has also participated in regional efforts at harmonizing these laws.

As an educator, she has participated in many outreach activities to key biotech stakeholders including policymakers and technocrats, parliamentarians, scientists, teachers, farmers and the media. These activities have covered areas spanning General Biotechnology, Risk Communication (Safety of GM Technology), Media Relations (for Scientists), the Art of

Reporting Science (for Journalists), Policy Formulations, and Regulatory Landscaping and Practice in Africa, with reference to adoption of GM Technology/Biosafety and Biosecurity.

Prof. Kenya is a founding member of the Programming Committee of the Open Forum for Agricultural Biotechnology in Africa, a flagship project of the African Agricultural Technology Foundation, headquartered in Nairobi, Kenya.

On the international scene, she has been very active in the activities of the Biological Weapons Convention (BWC) and the Organization for the Prevention of Chemical Weapons (OPCW) and is a member of the Chemical Weapons Convention Coalition. Her thrust in these organizations is to articulate the need for increased capacity strengthening and enactment of laws that will enable African countries gain from international efforts by the global community in making everywhere safe for human existence.

Tuğçe Kurtiş is a research affiliate of the Cultural Psychology Research Group at the University of Kansas. She completed her Ph.D. in Social Psychology and a graduate certificate in African Studies at the University of Kansas. Drawing upon perspectives in cultural, feminist, and critical psychologies and interdisciplinary discussions in decolonial studies and transnational feminisms, her research focuses on sociocultural constructions of subjectivity and relationality, which she examines through joint processes of voice and silence in interpersonal and collective experience. Her main objective as a social psychologist is to use psychological theory, pedagogy, and practice as resources for global social justice.

Ludwin E. Molina is an Associate Professor of Psychology at the University of Kansas. Broadly speaking, his research examines subgroup relations within diverse communities. This work spans topics of national identity, immigration, and respect.

Njiruh Paul Nthakanio holds a Ph.D. in Genetics and Plant Breeding (Zhejiang University, China). Formerly, I taught Molecular Biology, for a total period of over 20 years now, at Technical University of Kenya and currently at University of Embu (Kenya). At University of Embu, where I am Registrar Planning, Administration and Finance, I double up between administration and teaching. After a number of years practising science a spirit of reconnaissance seem to creep into me and I start questioning why the greater disparity in science and technology between

Africa and the rest of the world. In pursuit of this, the question of how science and technology became latent in Kenya and in the rest Africa in face of colonialism keep surfacing up. This book chapter explores the numerous opportunities that can lead to financial and wealth gains if indigenous science and technologies can be catalyzed from latency.

Marilyn Oladimeji, ECE, ExAT, MSW, MScs, Ph.D. As an Expressive Arts Therapist, I believe that engaging in various forms of interventions using art-making can sometimes be described as "conscious dreaming", a type of dreaming from which you can gain a feeling of being in control of your body, mind, and soul. Membership: Ontario Expressive Arts Therapy Association (OEATA) and the National Guild of Hypnotists (NGH).

Luis Gómez Ordóñez is a Professor of Psychology and Director of the School of Psychology at the Universidad Nacional Costa Rica, he has also been a Researcher at The Latin American Studies Institute, School of Sociology and Education in the same university. He is a member of the Decolonial Studies Group and the Collective of Liberation Psychology.

Ignacio Dobles Oropeza is Professor of Psychology at the University of Costa Rica, where he has been Director of the School of Psychology and the Institute of Psychological Research. His current work is centred in social, political and community psychology.

Valerie Robert, M.Ed. is a teacher by profession and an aspiring film-maker. She has a worked for various education groups and not for profits, and endeavors to move the process of decolonization forward both within herself and in her work.

Kimberly L. Todd is a Ph.D. candidate in Social Justice Education at the University of Toronto (OISE). Her research interests include education, decolonization, Indigenous epistemologies, dreaming, and spiritual knowledges. She is a Canadian teacher with experience teaching in South Korea, the United Arab Emirates and at a First Nations school in Canada. Kimberly is a member in good standing with the Ontario College of Teachers. She is also a curricular resource designer and has created educational resources for organizations like Amazon Watch and the David Suzuki Foundation. She is passionate about decolonization and education.

Njoki Nathani Wane Professor Njoki Wane is a recognized scholar in the areas of Black feminisms in Canada and Africa, African indigenous knowledges, African women and spirituality. One of her most recent publications is "Indigenous African Knowledge Production: Food Processing Practices Among Kenyan Rural Women". She has co-authored an Anti-racist training manual, *Equity in Practice: Transformational Training Resource*, with Larissa Cairncross, *Ruptures: Anti-Colonial & Anti-Racist Feminist Theorizing* with Jennifer Jagire and Zahra Murad, *A Handbook on African Traditional Healing Approaches & Research Practices* with Erica Neeganagwedin. She has also co-edited *Spirituality, Education & Society: An Integrated Approach with Energy Manyimo & Eric Ritskes* and *The Politics of Cultural Knowledge with Arlo Kempf and Marlon Simmons*. Prof. Wane headed the Office of Teaching Support in 2009–2012. She has been nominated TVO Best Lecturer, and is the recipient of the Harry Jerome Professional Excellence Award (2008) and of the African Women Achievement Award (2007). She is also a recipient of the prestigious David E. Hunt Award for Excellence in Graduate Education for 2016, University of Toronto and the President of Toronto Teaching Award, 2017. She has recently become the Chair of the Department of Social Justice Education at the Ontario Institute for Studies in Education, at the University of Toronto.

CHAPTER 1

Introduction: A Meeting of Decolonial Minds

Njoki Nathani Wane and Kimberly L. Todd

Abstract This chapter examines how decolonial scholars have engaged and shaped decolonial discourse and posits that this text extends these discourses by delving into the potentiality for decolonial pedagogy to reformulate and reconfigure colonial structures. It maps the chapters within this text and engages various sites of colonial oppression by cutting across fields, geographies and institutions in order to trace decolonial resistance. This chapter highlights the topics covered in this anthology which include state power, the psychological sciences, education and Indigenous technologies and provides a layout of the strategies, critiques and research on how to transform these sites of colonial oppression.

Keywords Decolonial pedagogy · Resistance · Colonial oppression · Transformation

N. N. Wane · K. L. Todd (✉)
Ontario Institute for Studies in Education, University of Toronto, Toronto, ON, Canada
e-mail: k.todd@mail.utoronto.ca

N. N. Wane
e-mail: njoki.wane@utoronto.ca

N. N. Wane and K. L. Todd (eds.), *Decolonial Pedagogy*,
https://doi.org/10.1007/978-3-030-01539-8_1

1

Decolonization, according to Joseph's blog (2017), "is …a long-term process involving the bureaucratic, cultural, linguistic and psychological divesting of colonial power" (p. 1). Joseph explains that "decolonization is about shifting the way Indigenous Peoples view themselves and the way non-Indigenous people view Indigenous Peoples" (p. 1). Our book interrogates the notion of decolonizing pedagogy and in particular educational institutions. The aim of the book is to capture the fluidity of the decolonizing discourse. In the last decade, decolonization as a practice, theory or debate has been written about and researched so much so that, if we are not careful, we might lose our agency and the very essence of this important scholarship. In 1986, Ngugi wa Thiong'o released his book: *Decolonising the Mind.* In this book Wa Thiong'o called us to look into how western education was a colonizing tool and how language, culture and religion were central to colonizing mission. Wa Thiong'o building on the works of Fanon, Cheikh Anta Diop was calling on the colonized people to decolonize their minds. In addition, others who also echoed Wa Thiong'o were Chinua Achebe, Albert Memmi, Wole Soyinka, Ashis Nandy, Aimé Césaire, Linda Tuhiwai Smith, just to name a few. All these authors provided excellent analyses of the destruction of cultural traditions, education and any form of social fabric through the colonial machinery. Many of them went a step further and offered suggestion on how to decolonize from the colonial master. Some spoke about the importance of relearning Indigenous languages (Thiong'o 1986), while Smith (1999) talked of research as a dirty word and how the colonizers had justified their colonial agenda through research. Albert Memmi (1965) paid attention to the relationship between the colonizer and the colonized. Many questioned how colonial systems disrupted all forms of institutions, both public and private. In this anthology, our focus is on decolonizing the pedagogy. Our debate in this anthology on decolonial pedagogy in many ways, mirrors the journey that many scholars, educators, activists and researchers have taken, in trying to make sense of this colonial machinery that has an identity of its own.

Decolonizing of any form is a long and central component of colonized subjects. The chapters in this anthology provide an excellent analysis of how people have resisted despite the disruption. *Decolonial Pedagogy: Examining Sites of Resistance, Resurgence, and Renewal* therefore takes up the question of how to decolonize systemic structures, institutions and educational systems that have emerged out of colonial logics. This book delves into areas of psychology, education, spatial analysis and Indigenous technologies in an effort to critically engage sites of ongoing colonial oppression and delve into the potential for decolonial ruptures and transformation. This anthology demonstrates the potential inherent in decolonial work to cut across fields of study, engage Indigenous knowledges and transform sites of oppression. It highlights the need to heal colonial wounds and revitalize knowledges that fall outside of the colonial paradigm. This book utilizes provocative and critical research that takes up issues of race, the shortfalls of empirical sciences, colonial education models and the need for a resurgence in Indigenous knowledges to usher in a new public sphere; a public sphere brought about by decolonization. This book is a testament of hope that places decolonization at the heart of our human community.

The book consists of nine chapters wherein various authors engage the reader with the significance of decolonial pedagogy to usher in transformative change in various colonial sites. Additionally, this book subverts western hegemony by affirming the value of Indigenous technologies and the need for new education models that are inherently decolonial. This introduction provides the overview of the book, while the conclusion charts the way forward in regards to decolonizing the public sphere.

Ardavan Eizadirad in "The University as a Neoliberal and Colonizing Institute: A Spatial Case Study Analysis of the Invisible Fence between York University and the Jane and Finch Neighbourhood in the City of Toronto" (Chapter 2) uses a spatial analysis to trace the growth, expansion and development of both the Jane and Finch area and York University. The work is situated in anti-racist and anti-colonial framework by unveiling the racial, colonial and class logics that demarcate the two spaces. The Jane and Finch neighbourhood characterized by poverty, racism and violence is contrasted with York University, an institution characterized by privilege, whiteness and mobility. Eizadirad unfolds the process by which the university as an instrument of the State partakes in white privilege, racism and neo-colonialism at the expense of the Jane and Finch neighbourhood through implementation of invisible borders, segregation and a systemic racial hierarchy.

Glenn Adams et al. in "Decolonizing Knowledge in Hegemonic and Psychological Science" (Chapter 3) delves into the need for a shedding of colonial epistemic violence as a starting point for true liberation. They chart how the western hegemony inherent in the psychological sciences is failing a majority of the population. In an effort to address this issue head on, Glenn Adams et al. highlight their findings on the most effective approaches to begin decolonizing the psychological sciences. They call to light the strengths, limitations and possibilities inherent in each approach. This chapter provides much promise for the way in which the psychological sciences can become decolonial in practical and viable ways.

Kimberly L. Todd and Valerie Robert in "Reviving the Spirit by Making the Case for Decolonial Curricula" (Chapter 4) explore the need for alternative decolonial curricula to disrupt the hegemonic and colonial narratives of state curricula. It chronicles the challenges of Todd and Robert as they move through teacher's colleges and classrooms butting up against colonial structures and systems that are inherently violent for both teachers and students alike. Todd and Robert demonstrate how the Cartesian separations are weaved within the school day and how these separations perpetuate deeply embedded disconnections from mind, body, soul and nature. The chapter posits that alternative decolonial curricula are needed to lay the groundwork for a decolonial revolution in schooling and act as a catalyst for this process. The purpose of decolonial curricula is to provide decolonial tools for teachers to aid in the rupturing of epistemic colonial barriers that are inherent in provincial curricula (in a Canadian context) and elsewhere. As such, decolonial curricula should be designed to work in conjunction with the provincial/state curriculum in the average classroom and to be the building block for teachers' lesson planning. Decolonial curricula are posited as an act of resistance and revival in the wake of colonial structures. Decolonial pedagogies need to flourish within schools to begin the process of casting off the hegemony of western knowledges.

Chizoba Imoka in "Training for 'Global Citizenship' but Local Irrelevance: The Case of An Upscale Nigerian Private Secondary School" (Chapter 5) uses a case study to unveil western hegemony within the secondary school system in Nigeria. This chapter demonstrates how the curriculum and pedagogy flows from a western Eurocentric canon and with this comes a loss of Nigerian knowledges, languages and values. Imoka unveils the violence that this schooling enacts on the students and the need for resistance against such hegemonies. At the heart of this chapter is a need for decolonial pedagogy and transformation that honours the lived experiences, histories and ancestries of the student body. In order to further this analysis, Imoka provides a resource in the form of a lesson plan for teachers and students to engage the questions embedded in this chapter in a critical and transformative manner.

Marilyn Oladimeji in "Using Arts-Based Learning as a Site of Critical Resistance" (Chapter 6) takes up art as a medium for transformation both personal and collective. Oladimeji posits that art, because of its ability to generate new ways of perceiving and engaging with the world, acts as a mode of resistance against colonial oppression and western hegemony. Indigenous knowledges are deeply generative, creative, innovative and attuned to the moment and art provides a medium for tapping into these ways of being. Oladimeji advocates for arts-based learning for the well-being of students and as decolonial pedagogy that ruptures the colonial oppressive norms of education.

Njoki Nathani Wane in "Awakening the Seed of Kenyan Women's Narratives on Food Production: A Glance at African Indigenous Technology" (Chapter 7) charts her own journey back to her village where she studied food production and preservation technologies under the guidance of female elders. Through this process, Wane learned the value of Indigenous technology that had been passed down ancestrally from one generation to the next. Through this decolonial pedagogy, Wane shares with the reader the necessity for a revival of Indigenous technologies that fosters the flourishing of the mind, body and soul of the community. This chapter also chronicles the change that colonialism, globalization and environmental degradation has had on both the land and food preservation. It centres the voices of the Elders whose wisdom points the reader in a direction towards decolonial revival and resurgence. Ultimately, this chapter is a call to action to reaffirm relationships to ancestry, land, food and ourselves despite the continual abuses of colonialism.

Njiruh Paul Nthakanio and Eucharia Kenya in the "Role of Latent Local Technologies and Innovation to Catapult Development in Kenya" (Chapter 8) discuss at great length the value inherent in herbs, medicines and food for nourishment, healing, health and economic well-being. These Indigenous technologies have been subjugated or co-opted by colonial powers. Nthakanio and Kenya highlight various Indigenous technologies whether they be for the purpose of cultivation, healing or food preservation that has been abandoned or over looked in lieu of western technologies. This chapter foregrounds the need for revitalization in both the local epistemic valuing of these knowledges and the utilization of them for the emotional, physical and economic well-being of Kenyan people.

In our concluding chapter entitled "Conclusion: The Way Forward" (Chapter 9), we chart a course towards a decolonial future. We recap the poignant contributions of our authors and delve into some of the themes they have engaged in a new way. We chronicle the destructive nature of colonialism on the level of the soul, the ongoing harm it produces on the mind and body and the decolonial healing process that can emerge at the level of spirit. We posit that decolonial transformation provides a bridge for communal and collective transformation advocating for a re-emergence of spiritual technologies, courage and community. Our conclusion is not an ending but a beginning. May this book provide insights, strategies and fortitude to help you continue on your own decolonial journey.

REFERENCES

Joseph, B. (2017, March 29). *A Brief Definition of Decolonization and Indigenization*. Retrieved April 28, 2018, from https://www.ictinc.ca/blog/a-brief-definition-of-decolonization-and-indigenization.

Memmi, A. (1965). *The Colonizer and the Colonized*. Boston: Beacon Press.

Smith, L. T. (1999). *Decolonizing Methodologies: Research and Indigenous Peoples*. Dunedin: Otago University Press.

Thiong'o, N. W. (1986). *Decolonising the Mind: The Politics of Language in African Literature*. London: J. Currey; Portsmouth: Heinemann.

PART I

Challenging State Power

The University as a Neoliberal and Colonizing Institute: A Spatial Case Study Analysis of the Invisible Fence Between York University and the Jane and Finch Neighbourhood in the City of Toronto

Ardavan Eizadirad

Abstract Jane and Finch is notoriously known in the City of Toronto as a high profile "Priority Neighbourhood" characterized by poverty, crime, and violence yet it is situated in close proximity to York University, a place of higher learning characterized by modernism, order,

A more extensive version of this chapter was initially published in the *Journal of Critical Race Inquiry* in March 2017. A preliminary draft of this chapter was presented at the OISE 15th Annual Dean's Graduate Student Research Conference on March 7, 2015, in Toronto, Canada. Thanks to Dr. Sherene Razack for introducing me to the concept of spatial analysis.

A. Eizadirad (✉)
Ontario Institute for Studies in Education, University of Toronto, Toronto, ON, Canada
e-mail: ardavan.eizadirad@mail.utoronto.ca

© The Author(s) 2018
N. N. Wane and K. L. Todd (eds.), *Decolonial Pedagogy*,
https://doi.org/10.1007/978-3-030-01539-8_2

9

multiculturalism, and innovation. Using a spatial analysis, the first half of the chapter traces the social, cultural, and historical development of Jane and Finch and York University, contrasting the rapid development and expansion of York University in relation to the slow growth and deteriorating living conditions of Jane and Finch. The second half of the chapter explores the racialization of physical and social differences between York University and Jane and Finch, particularly how interlocking systems of domination produce, maintain, and (re)produce an invisible fence that constitutes York University as a civilized space and Jane and Finch as a spectacle of violence and terror. Overall, this chapter argues that the university as an extension of the State participates in the neoliberal and colonizing project of constructing York University as a safe place of higher learning at the expense of social marginalization, stigmatization, and exclusion to the Jane and Finch community and Othering of its residents. By racialization of Jane and Finch and Othering of immigrants and visible minorities, York University masks its own violence and protects and privileges Whiteness.

Keywords Spatial analysis · York University · Jane and Finch · White privilege · Racialization · Othering · Colonization · Neoliberalism

INTRODUCTION TO SPATIAL ANALYSIS: INTERLOCKING SYSTEMS OF DOMINATION AND OPPRESSION

Spatial analysis can serve as a medium to examine how interlocking systems of domination and oppression operate dynamically and simultaneously, co-constituting one another through each other (Razack 2002a). It is through such vantage point that the myth of space as naturally evolving is debunked (Lefebvre 1991) and one can transition to view elite spaces such as the university as a spatial product socially, culturally, and politically produced within a web of unequal power relations benefiting colonizers at the expense of oppression and marginalization to the colonized. Spatial analysis as a tool functions to denaturalize and disrupt what often is accepted as "the way things have always been" through contextualization and historicizing, referred to by Razack (2002a) as "unmapping". Razack (2002a) points out that, "unmapping is intended to undermine the idea of white settler innocence -the notion that

European settlers merely settled and developed the land- and to uncover the ideologies and practices of conquest and domination" (p. 5).

The objective of this spatial case study analysis is to unmap (Razack 2002a), historicize, and contextualize the role of York University as an apparatus of the White settler society participating in practices of conquest and domination which (re)produce racial hierarchies and a socially stratified Canadian society. The first part of the chapter traces the social, cultural, and historical development of Jane and Finch and York University, contrasting the slow and deteriorating living conditions of Jane and Finch in relation to the rapid development and expansion of York University. The second half of the chapter explores the racialization of physical and social differences between York University and Jane and Finch with an emphasis on how interlocking systems of domination produce, maintain, and (re)produce the invisible fence and how this benefits specific social groups at the expense of marginalization to others.

BRIEF HISTORY OF JANE AND FINCH

Jane and Finch is a neighbourhood located in the former city of North York in northwestern Toronto, Ontario, Canada centred around the intersection of two arterial roads: Jane Street and Finch Avenue. The area is roughly bounded by Highway 400 to the west, Driftwood Avenue to the east, Grandravine Drive to the south, and Shoreham Drive to the north (Leon 2010; Narain 2012). Starting in 2005, the City of Toronto identified thirteen "Priority Neighbourhoods" to receive extra attention for the purpose of neighbourhood improvement in various capacities. In March 2014, the City of Toronto expanded the programme to include thirty-one identified neighbourhoods and renamed the programme from "Priority Neighbourhoods" to "Neighbourhood Improvement Areas" (City of Toronto 2016). According to the City of Toronto (2016) website, the thirty-one neighbourhoods were selected through the Toronto Strong Neighbourhoods Strategy 2020 which identified areas falling below the Neighbourhood Equity Score and hence requiring special attention. Since inception of these neighbourhood improvement initiatives, Jane and Finch has always been one of the neighbourhoods identified as requiring special attention whether as a "Priority Neighbourhood" or a "Neighbourhood Improvement Area" (City of Toronto 2016). According to the City of Toronto *Jane-Finch:*

Priority Area Profile (2008), the neighbourhood has "an approximate population of 80,150 living within an area span of twenty-one kilometre squared with an average population density of 3,817 persons per kilometre squared". The neighbourhood is characterized by unemployment, single-parent families, and high percentage of visible minorities which leads to Jane and Finch being a constant target of negative media depictions. Within dominant narratives in the media, the social problems of the neighbourhood are often blamed on its residents without much attention given to the systematic and structural conditions which historically have influenced the neighbourhood's trajectory of development leading up to its current conditions.

The major print document that is available and accessible which covers the history of Jane and Finch is called *From Longhouse to Highrise: Pioneering in Our Corner of North York* published in 1986 by Downsview Weston Action Community with assistance from York University's Community Relations Department. It traces the historical roots of Jane and Finch to roughly around 1400 where the land belonged to Indigenous Peoples living in longhouses that "were up to 120 feet long and 30 feet wide" (p. 5). In describing how the settlers obtained the land the document states,

> Eventually, our Indian pioneers were forced to leave their village as the land could no longer support them. The soil had slowly lost its fertility, the forests had thinned making firewood and building timber scarce and the forest wildlife disappeared. The Indians moved north and west to new areas of plenty. (p. 5)

This simplistic description does not outline how the Indigenous Peoples were displaced and their lands stolen, with violence used as a tool for conquest and domination. Rather, it paints the picture that Indigenous Peoples left because "the land could no longer support them" and then the European settlers arrived shortly after. These narratives mask the violence that was enacted on Indigenous Peoples and their communities. Ironically, after obtaining the land European settlers primarily used the land for farming up until the end of World War Two.

The next major change to the land occurred after World War Two. Throughout the nineteenth century, the Jane and Finch area underwent incremental growth through investments "in building of schools, churches, and the addition of a railway in 1853" (Richardson 2008, p. 2). As World War Two ended, Canada as a nation-in-the-making seized the opportunity to build a unique identity for itself by implementing less restrictive immigration policies and accepting immigrants from non-European countries. As a means of being able to provide living spaces for newly arrived immigrants entering the country, Canada began to invest in building affordable housing in the suburbs. As a result, in 1954 the federal and provincial government expropriated 600 acres of farmland in the Jane and Finch area for the intention of developing 3000 low-cost homes (Lovell 2011).

Jane and Finch underwent massive development by the Ontario Housing Commission in the 1960s to keep up with rapid rates of new-comers entering Canada (Narain 2012). Jane and Finch represented an ideal choice for many new immigrants due to low rent costs and rela-tively close proximity to the downtown core of the city. At the time, immigrants that were moving into the Jane and Finch area were pre-dominantly from West Indies, Asia, Africa, South America, and India (Richardson, p. 3). High-rise apartments and townhouses were built at a rapid rate. This linear style of hollow urban planning without much thought to the internal infrastructure of the neighbourhood lead to the population of Jane and Finch expanding "from 1,301 in 1961 to 33,030 in 1971" which included establishment of twenty-one high-rise apartment buildings in the neighbourhood (Downsview Weston Action Community 1986; Rigakos et al. 2004). Jane and Finch contin-ued its exponential growth in the 1970s and 1980s. During this period the majority of its residents were working-class immigrants and visible minorities (Narain 2012).

Jane and Finch continued its trend of becoming a popular place of residence for immigrants and racialized minorities. As this trend con-tinued, the neighbourhood began gaining "a reputation for ethnic con-flict, crime, and violence which continues to haunt it today" (Richardson 2008, p. 76). In *"Canada's Toughest Neighbourhood": Surveillance, Myth, and Orientalism*, Richardson (2008) argues that the continuous nega-tive representation of Jane and Finch produced an image of the neigh-bourhood as an "uncivilized" space infested with social problems (p. 77). What needs to be questioned is how is it that from all the neighbour-hoods in the City of Toronto, Jane and Finch became the city's most notorious neighbourhood associated with violence, crime, and gangs?

David Hulchanski's (2007) report *The Three Cities Within Toronto* provides the means to contextualize the development of Jane and Finch in comparison with other neighbourhoods in the City of Toronto. The study provides a comprehensive examination of income polarization among Toronto's neighbourhoods from 1970 to 2005 taking into consideration neighbourhood demographics. Findings indicate emergence of three distinct cities within Toronto based on income change. "City #1" makes up 20% of the city and it is generally found in the downtown core of the city in close proximity to the city's subway lines. The neighbourhoods under "City #1" are identified as predominantly high-income areas where the average individual income increased by 20% or more relative to the Toronto Census Metropolitan Area (CMA) average in 1970. "City #2" makes up 40% of the city and is characterized by middle-income neighbourhoods. Individual incomes in "City #2" remained the same having undergone an increase or decrease of less than 20%. "City #3" makes up 40% of the city and includes the Jane and Finch area. Individual incomes in "City #3" have undergone a decrease of 20% or more. Other than income differences, there are other major differences between "City #1" and "City #3" particularly in terms of the number of immigrants and visible minorities living in the areas. 82% of "City #1" is White compared to 34% of residents in "City #3". The percentage of foreign-born people in "City #1" declined from 35 to 28% between 1971 and 2006, whereas in "City #3" the number of immigrants increased drastically from 31% of the population in 1971 to 61% in 2006 (Hulchanski 2007, p. 11). These statistical trends support and confirm that predominantly beginning in the 1970s the majority of residents of Jane and Finch became immigrants and visible minorities. Currently 70.6% of the Jane and Finch population consists of visible minorities (City of Toronto 2008).

Hulchanski's data demonstrates drastic differences in long-term neighbourhood trends within Toronto, but more importantly it deconstructs the fallacy that neighbourhoods simply evolve "naturally" and by chance. Long-term trends from the findings indicate that investments and resources are distributed inequitably to different neighbourhoods in the City of Toronto; neighbourhoods composed of majority White residents are privileged at the expense of systemic neglecting of neighbourhoods composed of majority working-class immigrants and visible minorities. For Jane and Finch, the lack of investment in the infrastructure of the neighbourhood along with lack of accessibility to social programmes, employment, and resources resulted in "the most ripe of conditions for concentrated urban poverty to develop" (Rigakos et al. 2004, p. 17). Instead of the City of Toronto acknowledging that urban planning of Jane and Finch was a recipe for failure, the City of Toronto along with media narratives began placing blame of the social problems plaguing the neighbourhood on the residents of the area targeting them for social, cultural, and racial biases (Zaami 2012).

Brief History of York University

The building of York University occurred during a period of rapid expansion for Canadian post-secondary institutions coinciding with the maturing of the "baby boom" generation after World War Two (Unterman and McPhail 2008). Canada's population was increasing at a rapid rate after World War Two due to the "baby boom" generation maturing into young adults and more flexible immigration policies. A large percentage of the Canadian population was young and as a means of addressing their needs, the provincial government of Ontario invested in building a new university in Toronto to increase accessibility to post-secondary education. As a result, York was incorporated in March 1959 "with the biggest task ahead being the actual development of the University" (York University 2012b).

The province anticipated that the new university would establish itself slowly due to lack of credibility. Therefore, one of the conditions outlined by the legislation which approved the building of York University was the requirement for the university to be affiliated with the University of Toronto in its early stages "for a period of not less than four years and not more than eight years" (Ross 1992, p. 66). This was intended to assist York University in transitioning to establish itself with credibility and adequate scholastic standards. York University began its operation in the fall of 1960 as an affiliated college of the University of Toronto. Classes took place on University of Toronto campus downtown. In the fall of 1961, classes were moved to York's Glendon College Campus located near Bayview Avenue and Lawrence Avenue East (Unterman and McPhail 2008, p. 9). At this time, York Board of Governors began searching for an ideal location to build a new large campus in order to establish York as an independent university that would accommodate greater number of students.

At the time of its incorporation York Board of Governor's had the vision of building "a large, urban, multi-Faculty campus that would eventually accommodate 7,000 to 10,000 students" (York University 2012b). The Keele site, located near Jane and Finch, was selected as an ideal location because it was expected that "the movement of population in Toronto would be towards the northwest and it would be accessible to future main traffic arteries". York pursued in securing such location and as a result in 1962 the Province of Ontario gave York University 474 acres of previously expropriated land near the Jane and Finch area to start the initial construction and development of the Keele Campus (Unterman and McPhail, p. 9). The planning and construction took approximately three years to be completed. In 1965, York University ended its affiliation with University of Toronto and declared independence by opening up its Keele Campus. The location of the Keele Campus proved to be a good choice as enrolment of students increased at a rapid rate over the following years and the area became more accessible by cars and public transportation. What York University did not expect was the neighbourhood adjacent to it becoming known as the most notorious neighbourhood in the City of Toronto associated with poverty, crime, and violence.

THE INVISIBLE FENCE BETWEEN YORK UNIVERSITY AND JANE AND FINCH

The following sections, through a spatial analysis, explore the physical and socio-spatial differences that construct and (re)produce the "invisible fence" between what constitutes as York University and Jane and Finch. York University and Jane and Finch have had unique and significantly diverse trajectories in terms of their development yet they nearly occupy the same space. Jane and Finch has become known as a "Priority Neighbourhood" in the City of Toronto associated with poverty, crime, and violence whereas York University has expanded and established itself as the "third largest university in the country with a community of 55,000 students and 7,000 faculty and staff" (York University 2012b). As the examples of the racialized differences discussed will demonstrate, the "invisible fence" manifests itself in a dynamic and intentional manner involving inequitable power relations between the university and the neighbourhood. It is significant to emphasize that the invisible fence is constructed intentionally and not accidental in nature.

SOCIO-SPATIAL CHARACTERISTICS OF THE INVISIBLE FENCE

Settler colonialism constructs the dichotomy between the "civilized" and "uncivilized" with reference to how the land is occupied and used. The notion that Indigenous Peoples did not properly use and cultivate the land and hence had no title to it forms the basis of the "doctrine of discovery" which provides the foundational framework for legal claims to the land (Smith 2005, p. 56). As Wolfe (2006) points out in *Settler Colonialism and the Elimination of the Native*, the ideological justification for the dispossession of the land from the Indigenous Peoples was that settlers "would use the land better" (p. 389). Within this paradigm, settler colonialism becomes synonymous with modernity whereas Indigenous Peoples are seen as "pre-modern".

Narratives explaining how the land changed ownership is predominantly told from the perspective of White settlers in which certain fragments of history are purposefully forgotten, erased, and silenced particularly the extensive violence enacted on Indigenous Peoples as a means of displacing them and dispossessing their land. York University, as an extension of the State, directly participates in reproducing White settler narratives through knowledge production. A clear example is the involvement of York University in publishing the document *From Longhouse to Highrise: Pioneering in Our Corner of North York* which depicts Indigenous Peoples as "savages" who migrated to another location due to infertility of the land. The production of such narratives by York University serves two purposes: masks the dark history of a White settler Canadian society and masks the dark history of the university in its role as a colonizing institution. As a result, the university space is perceived as an ahistorical space distant from the atrocities and violence of the past.

Similar to colonial discourse which relies on binaries, there is a dualistic difference in terms of the physical appearance and architectural design of buildings between Jane and Finch and York University. Jane and Finch is primarily characterized by government housing which includes semi-detached houses and high-rise buildings with some having up to "427 units and reaching as high as 33 floors" (James 2012, p. 31). Most of the buildings in Jane and Finch are now more than forty years old and are in deteriorating conditions. In contrast, York University's architectural design and representation is aligned with a Eurocentric appearance outlined in the 1963 York University Master Plan which imposes specific restrictions in terms of what can be built and how it can be built. For example, the height of all structures within the campus is "to be restricted to a maximum height of 125 feet" (Unterman and McPhail 2008 p. 11). According to the *York University Cultural Heritage Assessment Report* (2008) the "designed built form was to project and maintain an urban feel through closely spaced buildings set amongst paved or planted quadrangles, after the example of old European universities and towns" (p. 10). The production of a Eurocentric appearance assists York University in constructing a unique identity associated with order and modernism.

While Jane and Finch has experienced minimal improvements in terms of its physical living conditions since the 1960s, York University has undergone massive growth at a rapid rate to improve its campus buildings and facilities. York University continues to expand by receiving funding and sponsorship from public and private sectors. By focusing exclusively on profits and the interests of the university, York University directly participates in creating a visual divide between what constitutes as York University and Jane and Finch. The university utilizes its numerous funds to maintain, improve, and invest in the creation of new buildings, faculties, and facilities as a means of improving the physical representation of the university space and what it can offer. Prime examples of recent big projects include the building of a forty-five million dollar Aviva Centre Stadium which opened in 2004 and yearly hosts a prestigious tennis tournament, building of a sport stadium which hosted track and field events as part of the 2015 Pan-American games in Toronto, as well as the on-going development of a direct subway line to the York University Campus (York University 2012c).

Urban Neoliberalism and Privatization of Spaces

The university operates as an apparatus of the White settler society where its expansion is guided by capitalism and profit. Property initiatives are managed by the York University Development Corporation (YUDC) which was incorporated in 1985 by York University. YUDC is responsible to "develop and manage innovative, community-focused and profitable real estate projects, and to strengthen partnerships with the university community's many stakeholders, our neighbours and all levels of government" (York University Development Corporation 2012). YUDC is primarily concerned with profit rather than strengthening partnerships with the surrounding community. YUDC utilizes urban neoliberalism tactics to privatize property on the York Campus as a means of maximizing profit for the university, which in effect excludes residents of the Jane and Finch neighbourhood from accessing the space and its resources.

Boudreau et al. (2009) explore the implications of neoliberalism in how it influences regulation of everyday life by pointing out,

> Urban neoliberalism refers to the *contradictory re-regulation* of everyday life in the city. Built on models of technologies of power developed in the previous era, the everyday now has become a tight place where individuals (divided by class, race, gender, etc.) are suspended in a web of control and opportunity, rights and responsibilities, further massification, and controlled isolation. (p. 29)

An example of urban neoliberalism was used by York University in 2002 where YUDC sold about 130 acres of land near the south boundary of the York campus to Tribute Communities "at a per-acre-cost of $450,000 dollars" for the creation of 845 private houses marketed as "The Village at York" (Saunders 2005b, p. 21). Interestingly, during sale negotiations between YUDC and Tribute Communities, Joseph Sorbara the Chair of the YUDC Board explicitly stated that "there would not be an opportunity in this proposal to develop subsidized housing and the plan is not to develop student housing although students might end up renting space in some of the residential homes" (Saunders 2005b, p. 21). Although York University consented to the use of its name by the developer of the property, it has no ownership and legal jurisdiction within the privately owned residential community composed of condominium townhouses, semi-detached, and detached homes built in sync with the modernistic architectural design of York University. York University agreed as one of the conditions of the land purchase that the home owners would get free access to the university's facilities including the fitness centre, indoor pool, gym, squash courts, and the outdoor tennis courts (Saunders 2005a, p. 3). The close proximity to public transportation and accessibility to numerous highways significantly added to the property values of the houses.

The Village at York private housing community on the York University Campus serves as a prime example of neoliberal urban planning where "neoliberalism endorses privatization of formerly public space in order to enhance capital accumulation and establish spaces that can easily be governed" (Narain, p. 60). The houses within The Village at York are marketed as upscale and are expensive to buy relative to properties in the Jane and Finch neighbourhood, at the time ranging from $279,990 to $458,990 dollars (Saunders 2005a, p. 3). With the Toronto housing market prices growing exponentially in the last few years, the homes in The Village at York are now valued and are for sale in the range of $700,000–$900,000 which are unaffordable for the residents of Jane and Finch relative to the neighbourhood's average after-tax income of $49,155 (City of Toronto 2008). As a result, housing becomes a medium for social exclusion contributing to gentrification because only the wealthy have purchasing power giving them access to the space and the perks that comes with ownership of the house. In contrast, many residents of Jane and Finch who are renters or residents living in high-rise apartments associated with government housing are socially excluded from The Village at York. Goldberg (1996) refers to this process as "spatial circumscription" where through privatization "the racial poor were simultaneously rendered peripheral in terms of urban location and marginalized in terms of power" (p. 188).

Intentional Racialization of Space

Through privatization everyday life is "re-regulated" (Boudreau et al. 2009) by imposing restrictions in terms of who gets access to the property and what can be done within that space. The interlocking dynamics of race and class lead to the confinement of living space for the poor and the working class in urban communities. Spaces such as The Village at York become owned by the affluent and the wealthy while simultaneously residents of Jane and Finch are excluded from entering the space because of the private status of the property. This contributes, within the dominant imagination, to the establishment of a sense of order and autonomy within the private space distant from the chaos of the public, "uncivilized" space of the Jane and Finch neighbourhood.

An initial process in gentrification and homogenizing of certain spaces is the representation of the space as desirable in order to attract the White and wealthy. York University directly participated in this process through the rebranding project which renamed Jane and Finch to University Heights. This was proposed in 2007 by Councillor Anthony Perruza of Ward 8 which includes the Jane and Finch community. Narain (2012) in *The Re-Branding Project: The Genealogy of Creating a Neoliberal Jane and Finch* points out that the rebranding process is a top-down initiative guided with neoliberal tactics that works "to marginalize and gentrify communities by promoting the privatization of public spaces" without consultation with the residents of the neighbourhood (p. 55). Narain (2012) goes on to contrast the positionality and power of politicians, government agencies, and economic stakeholders with that of local residents of the community, asking the important question "re-branding for whose benefits?". Narain demonstrates that although at the superficial level the rebranding of the neighbourhood might sound like a good idea as it aims to deter the politics of fear and danger that plague the Jane and Finch community, latently it supports gentrification by contributing to raising property values in the area by associating the neighbourhood with York University and the on-going construction of a direct subway line through the neighbourhood. This results in the exclusion of the poor.

What is more problematic about the rebranding of the Jane and Finch neighbourhood to University Heights is that it does not constructively attempt to resolve the systemic social issues plaguing the Jane and Finch neighbourhood. As a means of voicing their concerns, in protest of the name change, many key social organizations in the neighbourhood such as the *Jane and Finch Action Against Poverty* did not support the name change arguing that although York University is in proximal location to the neighbourhood it has not played a pivotal role in serving the interests of the community (Narain, p. 73). Yet without consultation with the Jane and Finch community, York University supported the rebranding project and aligned it with its 50th birthday celebrations. Ninety banners were made and placed on poles in the Jane and Finch neighbourhood costing $50,000 dollars paid in full by York University (Kim 2008). These banners refer to the neighbourhood as University Heights in an attempt to reconstruct a new image of the neighbourhood associated with modernity, order, and progress.

There is a dialectical relationship between physical space, social space, and the identity of bodies moving through the spaces (Lefebvre 1991; Razack 2002b). Physical space in its formation and texture not only determines who has access to the space, but it contributes to co-constituting what activities take place in that space. Jane and Finch often has an identity imposed upon it by external parties reflecting its limited power in the identity-making process. In contrast, York University as an institution has the power to actively participate and engage in identity-making practices by being in control of how the university space is governed, who gets to be associated with the university as an executive member, staff, or student, and what kinds of knowledge is (re)produced and disseminated through its faculties and staff. These differences in power relations result in race being positioned and interpreted differently in relation to space. Mohanram (1999) points out, "racial difference is also spatial difference, the inequitable power relationships between various spaces and places are rearticulated as the inequitable power relations between races" (p. 3). It is estimated that the residents of Jane and Finch are "from over 72 countries and speak about 120 different languages" (James 2012, p. 34) yet despite this biodiversity media coverage disproportionately features Black people in negative stereotypical portrayals of Jane and Finch. This process homogenizes what Jane and Finch represents and what it means to be Black. In contrast, when Blackness is associated with the university it is positioned in relation to discourses of multiculturalism. As "multiculturalism" is celebrated on the York University campus, what gets silenced and forgotten is how York University's decision-making continues to be dominated by White bodies who direct the university's funds and resources in serving the vested interests of a White settler Canadian society.

York University constructs an identity for itself by associating the university space with higher class and privilege. Bodies associated with York University, primarily faculty members and students, gain the ability to move through the invisible fence by conducting research in the Jane and Finch community and profiting from it through knowledge production and dissemination. This process resembles what Razack (2002b) calls "the racial journey into personhood" where White bodies "move from respectable space to degenerate space and back again" in an adventure that confirms they are in control and that they can survive a dangerous encounter with the racial Other (p. 127). In contrast, racialized bodies who are residents of the Jane and Finch community often do not get access to the university space, its facilities, and the knowledge produced and disseminated. These differences in terms of social mobility, in and through spaces, demonstrates how racialized bodies are marked as inferior and socially excluded from elite spaces of the university campus. Mohanram (1999) contrasts the mobility of White and Black bodies by stating, "Whiteness has the ability to move and the ability to move results in the unmarking of the body. In contrast, Blackness is signified through a marking and is always static and immobilizing" (p. 4).

POWER, POVERTY, AND UNIVERSITY–COMMUNITY PARTNERSHIPS

According to Murji and Solomos (2005), "as a concept racialization is useful for describing the processes by which racial meanings are attached to particular issues, often treated as social problems, and with the manner in which race appears to be a, or the, key factor in the way they are defined and understood" (p. 3). Within this framework, poverty is racialized and situated as born from Othered bodies from Jane and Finch. In contrast, York University positions itself as a sterile space free from poverty. Yet, poverty is not an independent variable but rather a result of unequal distribution of wealth and resources. Majority of Jane and finch residents who live in high-rise buildings in the neighbourhood rent as part of Toronto Community Housing initiative which seeks to provide affordable housing to a wide range of individuals primarily "seniors, families, singles, refugees, recent immigrants to Canada and people with special needs" (Toronto Community Housing 2012). The high number of high-rise buildings in the Jane and Finch community has implications in terms of the overcrowding it creates and the power struggle to access the limited resources available within the neighbourhood. In *Poverty by Postal Code 2: Vertical Poverty* (2011), United Way of Greater Toronto "tracks the continued growth of the spatial concentration of poverty in the City of Toronto" indicating that high-rise housing plays a role in this trend (p. ii). This is done through a longitudinal examination of the City of Toronto Census Data over a 25-year period from 1981 to 2006. One of the key findings of the report is that "poverty is becoming increasingly concentrated in high-rise buildings" (p. iv). The report identifies the construction of new private sector housing targeted exclusively for more affluent families, similar to the private housing created as at The Village at York, as a significant factor in the growing concentration of low-income tenants in high-rise buildings.

The Colour of Justice Network (2007) points out that "racialized communities experience ongoing, disproportionate levels of poverty". This is supported by the fact that "between 1980 and 2000, while the poverty rate for the non-racialized (i.e. European heritage) population fell by 28%, the poverty among racialized families rose by 361%" (The Colour of Justice Network 2007). The elimination of poverty in its entirety does not serve the interests of a White settler society because poverty provides the social conditions that make it possible to exploit racialized Others for labour (United Nations 1997). For colonizers, the question is not how to eliminate poverty but rather how to limit poverty to confines of certain spaces predominantly occupied by racialized bodies as a means of exploiting them for their labour (Goldberg 1996). Poverty functions as a required component for the smooth functioning of elite spaces such as York University, where the economically powerful can depend on those living in poverty to accept minimum wage jobs often without any benefits. Many immigrants are well-educated and have educational degrees from their country of origin, but upon arrival to Canada their degree is not recognized or undervalued by the Canadian government (Block and Galabuzi 2011, p. 4). For York University, the close proximity of the Jane and Finch neighbourhood provides them with access to a large pool of racialized immigrant bodies who are highly educated but willing to begin work in minimal paid jobs due to lack of accessibility to meaningful employment (City of Toronto 2008). National figures for Canada show that in the year 2000, immigrant men earned 63.1 cents for every dollar earned by a native-born Canadian with the same educational level (Boudreau et al., p. 91).

York University constructs its image as an innocent, pure space free of external challenges by emphasizing its role in having "experts" who assist the Jane and Finch community in finding solutions to social problems of the neighbourhood through university–community partnerships. York University has created various "partnerships" with the Jane and Finch community as a means of trying to make education more accessible for students living in the neighbourhood. One of the well-known partnerships that York University has with the Jane and Finch community is the Westview Partnership which originated in 1992. This partnership is currently associated with twenty-three schools in the Jane and Finch community and "seeks to enrich the schooling experience through programs designed to meet the needs, interests and expectations of students, teachers, parents, administrators and teacher candidates" (York University 2007; James 2012, p. 72). Students are often mentored within various programmes offered as part of the Westview Partnership and upon completion of the programme are provided with financial assistance or a university credit to facilitate their transition to start a post-secondary education at York University. A more recent partnership established with the community is a project called "Assets Coming Together for Youth: Linking Research, Policy, and Action for Positive Youth Development" which was approved for receiving one million dollars over five years between 2009 and 2014 from the Social Sciences and Humanities Research Council of Canada (York University 2013). The overall objective of the project is "to develop a comprehensive youth strategy that articulates how urban communities like the Jane and Finch community, can energize community assets that support positive youth development" (York University 2013).

Although these "partnerships" with the community seem like great ideas, in reality these initiatives are more about the university and the maintenance of a certain identity rather than the interests and the collective needs of the Jane and Finch community. That is not to say that these "partnerships" have no impact at all for some members from the Jane and Finch community. These "partnerships" do assist in making education more accessible financially for the selective few students who get to participate as part of the programme, but what remains problematic is that such initiatives do not address the systemic root causes of inaccessibly to education. The social conditions and infrastructures that form the systematic and structural barriers making it unlikely for youth from the Jane and Finch community to pursue post-secondary education remain unaltered, particularly once these partnerships cease to exist. Simply put, these initiatives invest in individuals and not systems. Although "one by one" and "student by student" such partnerships make a difference, unfortunately greater numbers of youth at a much rapid rate continue to become victims to the cycle of poverty and violence plaguing the neighbourhood (York University 2007).

SURVEILLANCE AND POLICING OF SPACES

York University carves a unique identity for the university space by emphasizing how it is different and distant from Jane and Finch. The problem is not that the spaces are different, but the consequences that arise from assigning social values to the differences. A binary is constructed marketing the university as a "civilized" space associated with Whiteness and multiculturalism. In contrast, Jane and Finch is portrayed as an "uncivilized" space through racialization of the neighbourhood, referring to the process by which Jane and Finch is labelled negatively based on a racial identity. As Frantz Fanon (1963) states in *Wretched of the Earth*, "the colonial world is a compartmentalized world" and "what divides this world is first and foremost what species, what race one belongs to" (pp. 3–5). Marking of Othered bodies becomes part of a colonial cycle which perpetuates unequal power relations by depicting two different species occupying York University and Jane and Finch. Immigrants and visible minorities are Othered and proclaimed different from Canadians by constantly being depicted in stereotypical representations such as perpetrators of violence. This leads to racialized Others being perceived as a threat, troublesome, and inherently violent.

One mechanism that contributes to establishing and maintaining a social divide between York University and Jane and Finch is the difference in policing of the space. This perpetuates the binary of two distinct worlds separated and occupied by two different species (Fanon 1963). York University primarily relies on closed circuit television cameras and 24 hours a day, 365 days-a-year security officers to maintain safety on its campus (York University 2012a). On the other hand, Jane and Finch is heavily surveilled and monitored by Toronto Police Services. The local 31 Police Division receives additional police officers to patrol the Jane and Finch neighbourhood as part of the Toronto Anti-Violence Intervention Strategy (TAVIS) which focuses on reducing guns and gang violence in "Priority Neighbourhoods" . TAVIS was launched in 2006 as a response to sharp increase in gun-related crimes in the City of Toronto. In 2005, 78 homicides occurred and majority of them in the summer months. The media labelled 2005 "summer of the gun" and "year of the gun" because 52 of the 78 homicides involved guns including the high profile death of fifteen-year-old White bystander Jane Creba on a busy downtown street on Boxing Day outside of the Eaton Centre shopping mall (James, p. 38). The killing of Creba dispelled the comfort that the violence can be contained within racialized spaces (James 2012). As a result, government funding was approved and provided to the Toronto Police Services for the launch of TAVIS, where additional police officers are "assigned to areas experiencing an increase in violent activity" using "crime trend analysis, occurrence of mapping, and community consultation" (Toronto Police Services 2012a).

The TAVIS initiative contributes to over policing of Jane and Finch which leads to on-going racial profiling and marginalization of the neighbourhood as a "hot spot" for crime and violence (Narain, p. 81). According to the Toronto Police Services, "the success of TAVIS is not based on the number of arrests made but on the reduction of crime, enhancement of public trust and confidence, and the building of relationships" (Toronto Police Services 2012b). Yet, TAVIS has profoundly failed in enhancing "public trust, confidence, and building of relationships" with Jane and Finch community members due to racial profiling of community members in encounters with the police. Racialized bodies are often subjected to police raids, random searches, and relentless questioning for appearing "suspicious". This process has damaged the image of the police and their relationship with the members of the community to the point where many residents of the Jane and Finch neighbourhood refuse to cooperate with the police.

Although York University is within the boundaries of the local 31 Police Division, heavy police presence is never visible on the university campus. York University found itself in a dilemma of having its campus safety questioned after a string of sexual assaults occurred on its campus originating in 2007 and continuing sporadically until now. Being under pressure to step up its security measures, York refuses to resort to the same techniques that govern and police Jane and Finch because it can lead to blurring of boundaries between what constitutes as York University and Jane and Finch. Instead, York maintains its social divide by investing in alternative surveillance mechanisms to reassure and comfort faculty and students of safety on campus. In an open forum discussion on safety held in November 2012 on the Keele Campus, the President of York University Mahmoud Shoukri emphasized that in 2011–2012 alone York University invested 10.2 million dollars on safety initiatives. Shoukri went on to outline the latest developments including having hired ten additional security officers, all personnel retrained and certified with enhanced roles and responsibilities enabling them to make a citizen's arrest, implementation of a new system where security concerns are emailed to students, staff, and faculty and posted to social media websites such as Facebook and Twitter, addition of 17 new exterior emergency phones, instalment of 70 LCD screens across the campus as a means of making information more accessible, enhancing of lighting on campus, and increases in campus shuttle services (York University 2012a). Interestingly, the implementation of increased police presence on campus is avoided at all costs. This approach facilitates York University in maintaining its image as a "civilized" space distant from the violence and crime associated with the Jane and Finch neighbourhood.

Conclusion

York University as an extension of the State participates in the neoliberal and colonizing project by constructing an identity for itself associated with order, modernism and progress at the expense of marginalization and stigmatization to Jane and Finch and Othering of its residents. As a result, an invisible fence is constructed between what constitutes as York University and Jane and Finch. The invisible fence is maintained and (re) produced intentionally through differences that bring together the racialization of social and physical characteristics of Jane and Finch. Physical differences include the appearance and architectural designs of the buildings; Jane and Finch is primarily characterized by government housing which includes semi-detached houses and high-rise buildings in deteriorating conditions whereas York University's architectural design and representation is aligned with a Eurocentric appearance which strictly restricts what can be built on its campus and how it can be built. Social differences between Jane and Finch and York University are constructed and maintained through who can access the housing available on campus as reinforced through implementation of neoliberal policies and tactics which promote privatization. The racialization of social differences are also made visible by how bodies are surveilled and policed within the spaces; Jane and Finch is predominantly monitored through heavy presence of police officers and special police units such as TAVIS, whereas York University avoids presence of police at all costs and instead invests in alternative safety measures such as investing in closed circuit television cameras and 24 hours a day, 365 days-a-year security officers to ensure safety on its campus. As a result of these physical and social differences, a dominant racialized narrative emerges as told from the perspective of colonizers, supported by neoliberal and colonizing institutional practices of York University, that depicts and constructs two distinct worlds separated and occupied by two different species (Fanon 1963). The clean and spacious environment of York University represents and symbolizes a safe place of higher learning associated with modernism, order, and multiculturalism. In opposition, the deteriorating and over-crowded conditions of Jane and Finch constitutes it as a lawless zone (Wacquant 2008) infested with social problems associated with crime, gangs, poverty, and violence caused by racialized bodies often Black, visible minority, immigrants, and residents of the Jane and Finch neighbourhood. Overall, York University as an apparatus of the White settler society, through its policies, practices and operation continuously racializes Jane and Finch through the construction of the invisible fence, in the process contributing to (re)producing a racial hierarchy and a socially stratified society where Whiteness is deemed invisible and retains a currency of privilege at the expense of marginalization and oppression to others particularly residents of the Jane and Finch community.

REFERENCES

Block, S., & Galabuzi, G. (2011). *Canada's Colour Coded Labour Market: The Gap for Racialized Workers*. Wellesley Institute and Canadian Centre for Policy Alternatives. Retrieved from http://www.wellesleyinstitute.com/wpcontent/uploads/2011/03/Colour_Coded_Labour_MarketFINAL.pdf.

Boudreau, J., Keil, R., & Young, D. (2009). *Changing Toronto: Governing Urban Neoliberalism*. Toronto, ON: University of Toronto Press.

City of Toronto. (2008). *Jane-Finch: Priority Area Profile*. Prepared by the Social Policy Analysis & Research Section in the Social Development, Finance, and Administration Division Source: Census 2006. Retrieved from http://www1.toronto.ca/city_of_toronto/social_development_finance__administration/files/pdf/area_janefinch_full.pdf.

City of Toronto. (2016). *NIA Profiles*. Retrieved from http://www1.toronto.ca/wps/portal/contentonly?vgnextoid=e0bc186e20ee0410VgnVC-M10000071d60f89RCRD.

Colour of Justice Network. (2007). *Colour of Poverty*. Retrieved from http://www.learningandviolence.net/lrnteach/material/PovertyFactSheets-aug07.pdf.

Downsview Weston Action Community. (1986). *From Longhouse to Highrise: Pioneering in Our Corner of North York*. North York: Downsview Weston Action Community.

Fanon, F. (1963). *The Wretched of the Earth*. Paris, France: Presence Africaine.

Goldberg, T. D. (1996). 'Polluting the Body Politic': Race and Urban Location. In *Racist Culture*. Cambridge, UK: Blackwell.

Hulchanski, J. D. (2007). *The Three Cities Within Toronto: Income Polarization Among Toronto's Neighbourhoods 1970–2005*. Retrieved from http://www.urbancentre.utoronto.ca/pdfs/curp/tnrn/Three-Cities-Within-Toronto-2010-Final.pdf.

James, C. (2012). *Life at the Intersection: Community, Class and Schooling*. Winnipeg, MB, Canada: Fernwood Publishing.

Kim, C. (2008). *Rebranding a Community*. Retrieved from http://www.insidetoronto.com/news-story/29508-rebranding-a-community/.

Lefebvre, H. (1991). *The Production of Space*. Malden, MA: Blackwell.

Leon, D. (2010). *Community Resource and Need Assessment Version 2009 Jane and Finch.* Toronto Centre for Community Learning & Development. Retrieved from http://test.tccld.org/wp-content/uploads/2012/11/JaneFinch_2009-10_CRNA.pdf.

Lovell, A. (2011). *An Overview of Development in Jane-Finch: 1950s to Present* (PPT). Assets Coming Together for Youth. Retrieved from http://www.yorku.ca/act/CBR/AnOverviewofDevelopmentJaneFinch_AlexanderLovell.pdf.

Mohanram, R. (1999). *Black Body: Women, Colonialism, and Space.* Minneapolis: University of Minnesota Press.

Murji, K., & Solomos, J. (2005). Racialization in Theory and Practice. In K. Murji & J. Solomos (Eds.), *Racialization: Studies in Theory and Practice.* New York: Oxford University Press.

Narain, S. (2012). The Re-Branding Project: The Genealogy of Creating a Neoliberal Jane and Finch. *Journal of Critical Race Inquiry, 2*(1), 54–94.

Razack, S. (2002a). When Place Becomes Race. In S. Razack (Ed.), *Race, Space and the Law: Unmapping a White Settler Society.* Toronto: Between the Lines.

Razack, S. (2002b). Gendered Racial Violence and Spatialized Justice: The Murder of Pamela George. In S. Razack (Ed.), *Race, Space and the Law: Unmapping a White Settler Society.* Toronto: Between the Lines.

Richardson, C. (2008). *"Canada's Toughest Neighbourhood": Surveillance, Myth, and Orientalism in Jane-Finch* (Unpublished Master's thesis). Faculty of Social Sciences, Brock University, St. Catharines, ON.

Rigakos, G., & Kwashie, F., & Bosanac, S. (2004). *The San Romanoway Community Revitalization Project: Interim Report.* Retrieved from https://docs.wixstatic.com/ugd/785722_51bbd96db4b84f9aa4c2d12070e3e617.pdf.

Ross, G., M. (1992). *The Way Must Be Tried: Memoirs of a University Man.* Toronto, ON: Stoddard Publishing.

Saunders, E. (2005a). *Appendix.* Retrieved from http://www.yorku.ca/mediar/special/2005/LandAppendicesFINAL.pdf.

Saunders, E. (2005b). *Review of Sale of Lands by York University to Tribute Homes in March 2002.* Retrieved from http://www.yorku.ca/mediar/YorkReport.pdf.

Smith, A. (2005). *Conquest. Sexual Violence and American Indian Genocide.* Boston: South End Press.

Toronto Community Housing. (2012). *About Us.* Retrieved from https://www.torontohousing.ca/about.

Toronto Police Services. (2012a). *SIU Clears Officers.* Retrieved from http://www.torontopolice.on.ca/modules.php?op=modload&name=News&file=article&sid=5185.

Toronto Police Services. (2012b). *TAVIS*. Retrieved from https://www.toronto.
    ca/311/knowledgebase/kb/docs/articles/special-purpose-bodies-and-ex-
    ternal-organizations/agencies,-boards,-commissions-and-corporations-abccs/
    service-and-program-operating-boards/toronto-police-service/toronto-an-
    ti-violence-intervention-strategy-tavis.html.
United Nations. (1997). *Human Development to Eradicate Poverty*. Retrieved
    from http://hdr.undp.org/sites/default/files/reports/258/hdr_1997_en_
    complete_nostats.pdf.
United Way of Greater Toronto. (2011). *Poverty by Postal Code 2: Vertical
    Poverty-Declining Income, Housing Quality and Community Life in Toronto's
    Inner Suburban High-Rise Apartments*. Retrieved from http://www.united-
    waytyr.com/document.doc?id=89.
Unterman, R., & McPhail, B. (2008). *Cultural Heritage Assessment Report:
    Cultural Heritage Landscapes-York University*. Toronto, ON: York University
    Development Corporation, York University. Retrieved from https://www1.
    toronto.ca/city_of_toronto/city_planning/community_planning/files/pdf/
    yorku_draft_cultheritage_asses_rep_mar08.pdf.
Wacquant, L. (2008). *Urban Outcasts: A Comparative Sociology of Advanced
    Marginality*. Cambridge, UK: Polity Press.
Wolfe, P. (2006). Settler Colonialism and the Elimination of the Native. *Journal
    of Genocide Research, 8*(4), 387–409.
York University. (2007). *'Reaching Back' Key to Westview Partnership's Success*.
    Retrieved from http://www.yorku.ca/yfile/archive/index.asp?Article=7661.
York University. (2012a). *President's Open Forum on Safety: Stronger Together
    Than Apart*. Retrieved from http://yfile.news.yorku.ca/2012/12/02/
    presidents-open-forum-on-safety-stronger-together-than-apart.
York University. (2012b). *York University's History*. Retrieved from http://
    about.yorku.ca/history/.
York University. (2012c). *Toronto 2015 Breaks Ground on New Athletics
    Stadium at York*. Retrieved from http://yfile.news.yorku.ca/2012/11/19/
    toronto-2015-breaks-ground-on-new-athletics-stadium-at-york-university.
York University. (2013). *$1-Million Grant Backs Research on Marginalized
    Urban Youth*. Retrieved from http://www.yorku.ca/yfile/archive/index.
    asp?Article=12497 and http://www.yorku.ca/act/about.html.
York University Development Corporation. (2012). Retrieved from http://
    www.yudc.ca.
Zaami, M. (2012). *Experiences of Socio-Spatial Exclusion Among Ghanaian
    Immigrant Youth in Toronto: A Case Study of the Jane and Finch Neighbourhood*
    (Unpublished Master's thesis). Western University, London, ON.

PART II

Psychological Sciences

CHAPTER 3

Decolonizing Knowledge in Hegemonic Psychological Science

Glenn Adams, Tuğçe Kurtiş, Luis Gómez,
Ludwin E. Molina and Ignacio Dobles

Abstract The authors describe three approaches evident in a collection of papers about "decolonizing psychological science" that they co-edited for the *Journal of Social and Political Psychology*. In the *indigenous resistance* approach, psychologists draw upon local knowledge to modify hegemonic practice and to produce psychologies that better resonate with local realities. In the *accompaniment* approach, global experts from hegemonic centers work alongside inhabitants of marginalized communities in struggles for social justice. In the

G. Adams (✉) · T. Kurtiş · L. E. Molina
University of Kansas, Lawrence, KS, USA
e-mail: adamsg@ku.edu

L. E. Molina
e-mail: ludwin@ku.edu

L. Gómez
National University of Costa Rica, Heredia, Costa Rica

I. Dobles
University of Costa Rica, Heredia, Costa Rica

denaturalization approach, psychologists draw upon local knowledge of marginalized communities to illuminate and resist the epistemic violence inherent in standard forms of hegemonic psychology. Together, these approaches extend consideration beyond the production of local psychologies attuned to the conditions of particular communities and illuminate decolonial versions of global psychology for general application.

Keywords Accompaniment · Cultural psychology · Decolonial theory · Denaturalization · Indigenous psychology · Liberation psychology

Throughout his life, revolutionary psychiatrist Franz Fanon witnessed firsthand the material violence of European colonial rule (see Bulhan 1985).[1] Fanon grew up as a subject of the French colonial empire in the Caribbean island of Martinique. He fought for the French against the Nazis in the Second World War. He faced racism as a person of African descent navigating the professional academy in postwar France. He treated victims and perpetrators of torture as a psychiatrist working for French colonial forces in Algeria. He fought alongside comrades in the Algerian liberation movement. These experiences provided Fanon with a standpoint from which to understand the Eurocentric modern order not as the leading edge of progress (as psychologists and other modern intellectuals typically imagine it), but instead as a product of the colonial violence that constitutes the "dark side" of modernity (Mignolo 2011). Beyond physical violence, Fanon's unique contribution was to illuminate psychological and epistemic manifestations of colonial violence that other observers often failed to note. As a response to such epistemic violence, Fanon warned readers "not [to] pay tribute to Europe by creating states, institutions and societies which draw their inspiration from her" (Fanon 1963, p. 314). Instead, he emphasized that the task of mental decolonization required new models. "For Europe, for ourselves and for humanity, comrades, we must turn over a new leaf, we must work out new concepts, and try to set afoot a new [hu]man" (Fanon 1963, p. 315).

  With respect to psychological and epistemic violence among the colonized, Fanon was an acute observer of the ways in which colonial situations leave psychological traces that persist after formal colonial rule (see also Duran and Duran 1995). He was among the first to write about psychological manifestations of neo-colonialism in the postindependence era (Fanon 1963, 1967). His writing emphasized that the violence of colonialism is not only limited to occupation of land and material resources but is also extended to the occupation of mind and being (Bulhan 2015). One articulation of this idea in the discipline of psychology is the notion of *colonial mentality*: "internalized oppression, characterized by a perception of ethnic or cultural inferiority ... that involves an automatic and uncritical rejection of [colonized ways of being] and uncritical preference for [colonizer ways of being]" (David and Okazaki 2006, p. 241; cf. Fanon 1967; Memmi 1965). However, Fanon's work also emphasized that the experience of mental colonization was not limited to the inferiority complexes and collective self-hatred of colonial mentality, but extended to what one might call the *coloniality* (Mignolo 2011; Escobar 2007) of everyday life: displacement of local understandings or ways of being and replacement by colonizer understandings or ways of being. Even in cases where people emerged from colonialism with collective pride and self-respect intact, colonial occupation was associated with insidious forms of mental colonization—for example, language (Ngũgĩ 1986) and religion (Comaroff and Comaroff 1991)—that led even many revolutionaries to experience basic realities on colonizers' terms.

As a result, Fanon emphasized that the task of liberation from colonial oppression required not only decolonization of land but also decolonization of the spirit or mind as the first territory of resistance, liberation, and hope. For Fanon, the task of mental decolonization was not primarily an abstract, theoretical concern; instead, it was a necessary component in the process of political *and* existential liberation whereby the "wretched of the earth" (Fanon 1963) take direction of their lives, bodies, and spirits. For Fanon and other theorists of the decolonial turn, the point is not merely to give an account of the colonial condition, but instead to draw upon the resistance and wisdom of the oppressed in pursuit of global justice.

Besides extending the focus on colonial violence beyond material manifestations to consider psychological and epistemic manifestations, Fanon's work is also remarkable for extending consideration beyond experience of the colonized to investigate implications of colonial violence for psychological experience in general. In particular, Fanon (1967) was an early critic of what later writers have referred to *The Coloniality of Knowledge* (e.g., Lander 2000) and *The Coloniality of Being* (e.g., Maldonado-Torres 2007) in the modern global order. He recognized that modal patterns of life in colonizer societies are not, as mainstream scientific perspectives are inclined to portray them, the pinnacle of human progress or unremarked standards in social science and global policy (e.g., of economic development or human rights). Rather, these patterns developed in attunement to the violence and rapacity demanded by the exercise of colonial domination. From this perspective, decolonization of spirit and mind requires new concepts and ways of being—not only among the colonized or dominated but also among the colonizer or people in positions of dominance—that better reflect (and promote) the hopes and aspirations of broader humanity. Similarly, in the context of intellectual pursuits, the task of decolonial liberation requires new epistemologies and methods that draw upon the experience of people in marginalized or colonial spaces as a privileged standpoint from which to better comprehend the human situation.

Despite its influence and foundational status in the interdisciplinary fields of postcolonial, decolonial, or subaltern studies, discussions of Fanon's work are virtually absent from his home discipline of psychiatry/psychology.[2] To the extent that academic psychologists are aware of the work, they are inclined to discredit it as political activism outside the bounds of proper scientific activity. Few psychologists emphasize the extent to which the modern global order is an inherently racist enterprise born of colonial violence, and even fewer consider the implications of this everyday coloniality for theory, method, and standard knowledge in psychological science. Instead, hegemonic versions of psychological science[3] typically reproduce and augment this violence by functioning as a tool for the modern global order. This hegemonic vision holds up modern individualist ways of being as the self-evident pinnacle of cultural evolution, the leading edge of human progress, and an optimal prescription for the pursuit of happiness that the field takes to be synonymous with well-being (for critiques, see Adams et al. 2018; Becker and Marecek 2008).

In contrast, the common thread that underlies our work is a set of two related concerns. The first concern is that knowledge production in psychological science reflects and promotes the interests of a privileged minority of people in WEIRD[4] settings. The second concern is that colonial power has imposed these WEIRD ways of knowing and being as a hegemonic standard for universal application (including in settings of the "Majority World;" Kağıtçıbaşı 1995). These concerns led us to propose a special thematic section (STS) of the *Journal of Social and Political Psychology* devoted to the topic of "Decolonizing Psychological Science." In this chapter, we briefly discuss the theoretical foundations that inform our work and their relationship to indigenous knowledge perspectives. We then present an overview of the submissions to the STS, with particular attention to decolonial strategies that emerged across contributions.

Theoretical Foundations

The idea for the STS on Decolonizing Psychological Science arose from a series of collaborative seminars between psychologists from the Cultural Psychology Research Group (University of Kansas) and the Costa Rican Liberation Psychology Collective (Universidad de Costa Rica and Universidad Nacional de Costa Rica). The purpose of the seminars was to explore the intersection of theoretical perspectives (i.e., cultural psychology and liberation psychology) that define our respective groups.

Cultural Psychology

As various critics have noted, knowledge in hegemonic psychological science tends to have its base in research with participants from WEIRD settings (Henrich et al. 2010; cf. Arnett 2008). As a result of this narrow focus, researchers in hegemonic traditions of psychological science have extensively documented particular patterns of being associated with the sense of abstraction from context and freedom from constraint that Markus and Kitayama (1991) referred to as an independent self-construal. However, rather than understand these patterns as a particular phenomenon (rooted in neoliberal individualism and Eurocentric modernity), the universalist inclination of hegemonic psychology is to elevate these particulars to the status of natural standard that marks the leading edge of progress in human development. When researchers in this tradition do occasionally step outside WEIRD settings to conduct research in Majority World settings, they often encounter Other patterns that resonate with more embedded or sustainability-oriented ways of being. However, they tend to interpret these patterns through a colonial lens not as equally viable alternatives, but instead in Orientalist

(Said 1978) fashion as evidence of backwardness, ignorance, or some other deviation from normal (Okazaki et al. 2008).[5]

In contrast to the context-independent outlook of hegemonic psychological science, approaches to cultural psychology generally consider the idea of "mind in context": that is, how a person's habits of mind (i.e., thinking, feeling, desire) exist in a dynamic relationship of mutual constitution with the culturally evolved affordances that make up everyday human ecologies. From this perspective, the structure of mind exists not only in inherited genetic code or brain architecture but also in the culturally evolved tools for thinking and feeling deposited in the stuff of everyday worlds. Rather than a universal set of tendencies moderated by ecological contingencies, cultural psychology perspectives emphasize that much of what we take to be just natural are instead particular habits of being developed in attunement with local conditions.

The relevance of cultural psychology approaches for decolonizing psychological science lies in two decolonial strategies. The first decolonial strategy is to normalize Other patterns that hegemonic psychological science portrays as "abnormal." This strategy draws upon context-sensitive accompaniment with people in the course of everyday life to arrive at a more adequate, experience-near account of Other experience. Rather than pathological deviations from hegemonic standard or evidence of lag along a unidimensional trajectory of development, the normalizing strategy of a cultural psychology analysis directs scholars and students to consider the value (and validity) of Other ways of being.

The second decolonial strategy is to denaturalize the patterns that hegemonic psychological science elevates as natural standards. Whereas the first (normalizing) strategy is similar to mainstream accounts in its focus on Other ways of being as the phenomena that require explanation, this second (denaturalizing) strategy directs scientific observers to turn the analytic lens (Adams and Salter 2007) and rethink hegemonic standard knowledge from Other epistemic standpoints. The epistemic perspective of people in marginalized settings provides valuable resources for thinking outside established norms and seeing hegemonic standards as problematic in their own right. In particular, this perspective affords an understanding of hegemonic standards not so much as the frontier of human potential in societies on the leading edge of social development, but instead as epistemic violence made possible by racialized colonial domination.

LIBERATION PSYCHOLOGY

Perspectives of liberation psychology have emerged as one of the most influential perspectives of Latin American psychology (e.g., Barrero 2012; Burton and Kagan 2005; Dobles et al. 2007; Martín-Baró 1986; Montero 2007). Increasingly, a substantial body of work appears under the liberation psychology label from a wide variety of settings across the globe (e.g., Enriquez 1992; for reviews, see Burton and Gómez 2015; Comas-Díaz et al. 1998; Lacerda and Dobles 2015; Lykes and Moane 2009; Montero and Sonn 2009; Watkins and Shulman 2008).

Rather than an intellectual foundation in imported varieties of hegemonic psychological science, perspectives of liberation psychology have an intellectual foundation in emerging traditions of Latin American thought. Whereas the former are attuned to everyday life in the modern (and Eurocentric) global order, the latter are attuned to everyday life in Latin American settings—realities that are the ongoing legacy of colonial violence. Foremost among these Latin American thought traditions are perspectives of liberation theology (e.g., Dussel 1972; Gutiérrez 1971) and especially the idea of a *preferential option for the poor*: that is, a commitment to treat the perspectives and interests of oppressed and impoverished communities as a primary source of insight about everyday truth and social reality. An equally important intellectual foundation from Latin American thought traditions is a participatory action orientation in pedagogy (e.g., Freire [1970] 1993) and social science (e.g., Fals Borda 1987) that emphasizes community-engaged praxis over insulated theory.

The relevance of liberation psychology perspectives for the decolonial project concerns the commitment to a practical science with and for the oppressed global majority, rather than an intellectual formation that reflects and reproduces the interests of the Eurocentric global order (and its local manifestations). Its key intellectual or epistemic orientation is a deliberate attempt to understand reality from the perspective of the oppressed. A liberation psychology perspective emphasizes that conventional representations of everyday events and experience are neither the neutral reflection of objective truth nor a "natural" reality. Instead they represent particular constructions of reality that reflect and serve the interests of the powerful. From this perspective, researchers can contribute to the decolonial project by conducting research that draws upon insights and repressed knowledge traditions of people in marginalized settings as resources for de-ideologization (Martín-Baró 1986) to expose and counteract the constructed, ideological nature of supposedly neutral or natural realities.

INDIGENOUS KNOWLEDGE

Indigenous knowledge perspectives have a similar role in both cultural and liberation psychology. The interest in indigenous knowledge is not to uncover the authentic body of traditional knowledge for some reified cultural entity—what one might refer to as *Indigenous* (with a capital *I*). Instead, the interest is indigenous (with a small *i*) knowledge as context-engaged, local understanding. Perspectives of liberation and cultural psychology draw upon such local knowledge as a source of important insight about everyday reality. Likewise, the insights that indigenous knowledge yields are not simply a more adequate psychology for local application. Instead, local knowledge provides a privileged epistemological standpoint from which to de-ideologize and/or counteract hegemonic psychological science in general.

DECOLONIAL STRATEGIES

One can productively think about the theme of decolonizing psychological science in two senses. The first sense of "decolonizing" refers to a process that one applies to standard knowledge and practice of science. In this sense, the task of decolonization requires that researchers reveal and counteract the colonial standpoint of standard scientific forms that typically masquerade as position-less or politically innocent reflections on objective reality.

The second sense of "decolonizing" refers to the production of knowledge practices suitable for the task of decolonization. In this sense, the task of decolonization requires the development of alternative concepts and tools that provide a broader foundation for the study of humanity. An important resource for this task is the work of critical scholars who operate from supposedly "peripheral" fields of knowledge production such as indigenous studies, ethnic studies, transnational feminism, Critical Race Theory, Decolonial Theory perspectives of Latin American Studies, Subaltern Theory perspectives of South Asian Studies, Postcolonial Theory perspectives in African Studies, and critiques of normativity in Disability Studies and Sexuality Studies. In contrast to the prevailing academic construction of these fields as sites for the application of general/central knowledge to particular/peripheral cases, the decolonial project requires something akin to *Theory from the South* (Comaroff and Comaroff 2012; cf. Connell 2007; Santos 2014). Specifically, it requires that researchers consider the epistemological perspective of geographically and socio-politically marginalized positions as a privileged source of general knowledge for the mainstream academic enterprise.

When we proposed a STS on decolonizing psychological science, our goal was to provoke consideration of the topic and to get some initial sense for how people understood it. In order to provide some structure for contributions to the STS, we posed a series of guiding questions in a supplement to the call for papers. One set of questions considered psychological consequences of colonial domination. What is colonial mentality, and how does one oppose it? To what extent is the modern global order—and the "individualist" varieties of psychological experience considered to be standard in the modern global order—a manifestation of ongoing colonial violence? A second set of questions considered whether there might be a place for psychology in the broader set of fields that consider decolonization and global social justice. How can psychologists help to decolonize psychological science and other global institutions (e.g., economic order, human rights regimes, international development practice)? Does psychology offer tools for decolonizing knowledge/consciousness? Can research within marginalized communities illuminate alternative bases for a psychological science that promotes sustainable well-being for global humanity? What is the appropriate balance of theory and praxis for a decolonial psychology? A third set of questions considered the extent to which the discipline of psychology—especially in the form of hegemonic psychological science—is itself a colonial form rooted in European global domination. (How) do theory, research, and practice in mainstream psychology contribute to oppression and constriction of life chances among people in marginalized communities? In what sense does psychological science carry the residue of past domination and reproduce present domination? In what ways do theory, research, and practice of mainstream psychology (and related disciplines) reflect and promote interests of power and privilege? This long list of questions reflected our desire to solicit a broad range of perspectives on the task of intellectual decolonization.

In response to the call for papers, we received 39 proposals from diverse geographic locations, including Aotearoa//New Zealand,

Australia, Brazil, Central America, Cuba, Greece, Guam, Hawaii, India, Indigenous in USA, Israel, Lebanon, Malaysia, Mexico, Palestine, Philippines, Somaliland, South Africa, Turkey, the United Kingdom, and people of African descent in the United States. These proposals reflected a diverse variety of intellectual perspectives, including community psychology, critical psychology, cultural psychology, disability studies, experimental social psychology, feminist studies, history of psychology, liberation psychology, participatory action research, queer studies, and social work. We ultimately selected eight articles for publication in the STS. In our review of these submissions, we discerned three decolonial approaches that we describe briefly in the following sections. (For a lengthier discussion, please see the introduction article to the STS; Adams et al. 2015.) Although the approaches often overlapped in practice, we present them as separate strategies for purposes of analysis.

INDIGENOUS RESISTANCE

The most prominent decolonial approach in submissions for the STS was what we refer to as *indigenous resistance* (Dudgeon and Walker 2015). In this approach, locally grounded researchers and practitioners reclaim local or indigenous wisdom to produce forms of knowledge that resonate with local realities and better serve local communities. The particular strength of *indigenous resistance* approaches is the valorization of local understanding as a legitimate source of knowledge. The tendency in hegemonic psychological science is to treat indigenous understanding as mere folk belief rather than reality-tuned knowledge based on empirical observation. In response, *indigenous resistance* approaches enact the decolonial strategy of *normalization*: treating as healthy or legitimate the indigenous understandings that hegemonic science devalues as unhealthy or illegitimate. Related to the valorization of local understanding, another strength of indigenous resistance approaches is to counteract not only tendencies of colonial mentality (David and Okazaki 2006) but also other forms of epistemic violence—especially the neoliberal individualist ways of being that colonial power has elevated to the level of hegemonic standard. Indigenous understandings provide a potential antidote to this epistemic violence by illuminating forms of intervention and ways of being that are better suited to local history and ecological conditions.

A potential limitation of indigenous resistance approaches lies in tendencies of what one might call romantic conservatism, whereby researchers or practitioners understand local ways of being as vestiges of timeless tradition rather than ecologically responsive patterns. This romantic conservatism becomes problematic when it leads people to interpret dynamic adaptation as rejection of cultural identity and inauthentic assimilation. Even more problematic are varieties of romantic conservatism that lead people to act as apologist for harmful practices in the name of tradition and avoid subjecting local understandings to critical scrutiny.

ACCOMPANIMENT

The second decolonial approach that we observed in submissions to the STS is *accompaniment* (Segalo et al. 2015; Watkins 2015). In this approach, researchers and practitioners leave the gated enclaves of Eurocentric global modernity and travel to colonized or racially subordinated settings to work and learn alongside people from marginalized communities in their struggles for social justice (Tomlinson and Lipsitz 2013; Watkins 2015). The decolonial force of these approaches lies in disrupting prevailing ideologies of scientism (e.g., Denzin and Lincoln 2012, 2018) that mandate cool detachment from societal struggles as the preferred mode of intellectual inquiry. In contrast, accompaniment approaches to the decolonial project suggest that one comes closest to truth when one participates alongside marginalized communities in the context of everyday struggles. Accordingly, one strength that accompaniment approaches share with indigenous resistance approaches is an emphasis on place-based methods and resulting knowledge (Tuck and McKenzie 2014) that directs attention from concerns that dominate hegemonic psychological science to the concerns of people in marginalized communities of the Majority World. An additional strength of accompaniment approaches is to demand involvement of psychological scientists who live and work in the WEIRD settings, whose position of privilege should not relieve them of responsibility to pursue decolonial justice.

Serious limitations of these approaches are related to the position of the accompanying researcher or practitioner as an outside expert. The expert's position of relative power and privilege can enable such colonial practices such as *epistemic extractivism* (Grosfoguel 2016; see Smith 1999) whereby experts stay briefly in marginalized communities and appropriate local voices for their own professional agendas.

A related issue is what Cole (2012) referred to as "The White-Savior Industrial Complex": a belief that solution of global injustice requires the intervention of heroic outsiders who will lead oppressed Others to liberation. The problematic feature of this complex is not merely its relation to paternalistic colonial narratives, but also that the focus on saving others obscures experts' own participation in enduring forms of colonial privilege and systemic domination from which they benefit.

A less obvious limitation of accompaniment approaches—one that they share with indigenous resistance approaches—is an understanding of decolonial work as something that one does in colonized or marginalized settings of the Majority World outside the affluent spaces that most psychologists inhabit. Even in exemplary applications, the focus of these approaches is typically on giving psychology away to counteract acute, political, and economic manifestations of colonial violence in sites of oppression. These approaches typically do not turn the analytic lens to consider how hegemonic psychological science reflects and reproduces more diffuse and enduring forms of colonial violence.

DENATURALIZATION

This exclusive focus on colonized settings contrasts with a third decolonial approach, which we refer to as *denaturalization* (Kurtiş and Adams 2015; Phillips et al. 2015). The denaturalization approach suggests that decolonial efforts are incomplete if they focus only on consequences of violence among the colonized. In addition, these efforts must also illuminate and disrupt elements of coloniality in standard regimes of hegemonic science (i.e., *The Coloniality of Knowledge*) and the psychological habits of the people in the typically WEIRD settings that inform scientific imagination (i.e., *The Coloniality of Being*).

Along these lines, the primary strength of denaturalization approaches is to illuminate *The Coloniality of Knowledge* and being in standard patterns of hegemonic psychological science. When psychologists prescribe color-blind modes of perception that afford denial of racist oppression (Phillips et al. 2015) or neoliberal individualist models of relationality that afford narrow investments of care at the expense of broader community (Kurtiş and Adams 2015), they reproduce acts of epistemic violence. This violence is the result not merely of imposing models that are inappropriate for local realities, but also of elevating particularly positioned tendencies to the level of white-washed standard. The primary goal of the denaturalization approach is to counteract these forms of epistemic violence and to illuminate alternative ways of being—critical consciousness of racism versus color-blind ignorance, sustainable relationality versus the pursuit of growth—that better serve interests of humanity. The decolonial potential of these alternative ways of being is not only to better reflect the epistemic perspectives of people in racially oppressed communities but also to provide a more just and sustainable foundation for a global psychology. In other words, the radical goal of this approach is not (only) local liberation but (also) to promote decolonial versions of knowledge and practice that answer Fanon's call to develop new concepts for a more human(e) science.

LIMITATIONS

The primary strength of denaturalization approaches is also their primary limitation. The audience of these approaches is hegemonic psychological science rather than people living in situations of colonial and racial oppression. Similarly, the proximal target of decolonial efforts is forms of knowledge, not communities of people. For researchers who accompany people from marginalized communities in struggles against material violence, this focus on epistemic violence from the insulated security of academic institutions risks becoming the very sort of sanitized, intellectual exercise that requires decolonial intervention.

THINKING THROUGH INDIGENOUS KNOWLEDGE

As we stated at the outset, our discussion of decolonial approaches as separate categories has been somewhat artificial. Rather than mutually exclusive categories, successful efforts to decolonize a field of knowledge are likely to incorporate elements of each approach. Although accompaniment or indigenous resistance approaches properly emphasize local understanding, they are most effective when they draw upon this resource to turn the analytic lens and counteract *The Coloniality of Knowledge* in hegemonic psychological science. Similarly, denaturalization approaches are typically most effective when they take inspiration from indigenous knowledge or a history of engagement with everyday life in marginalized settings.

The unifying theme across these decolonial approaches is the value of engagement with indigenous or local knowledge, whether through accompaniment in the course of everyday life (Martín-Baró 1986; Watkins 2015) or through training in the perspectives of place-based (Tuck and McKenzie 2014) or identity-conscious epistemologies (e.g., Critical Race Theory; Crenshaw et al. 1995).

The point of engagement with these indigenous knowledge formations is not for romantic celebration of multicultural diversity, but instead what one might (to paraphrase Shweder 1990) refer to as *thinking through indigenous knowledge*. In one direction, epistemic perspectives of indigenous communities help to illuminate *The Coloniality of Knowledge* and being. They challenge the epistemic privilege of WEIRD spaces and disrupt the epistemologies of ignorance (Mills 2007) that enable discussions of history and society without reference to colonial violence. In the other direction, epistemic perspectives of indigenous communities provide researchers and practitioners with models of personhood and social relations that better serve the interests of the broader humanity. They may be productive of sustainable well-being not only in indigenous and other marginalized communities but also as models for emulation in the WEIRD settings that inform hegemonic psychological science. In these ways, the epistemological standpoint of indigenous knowledge provides resources for perception that enable researchers and practitioners to notice and respond to everyday forms of coloniality and epistemic violence that typically remain hidden in plain sight.

NOTES

1. This chapter is an adaptation and selective elaboration of our introduction article in the special thematic section on "Decolonizing Psychological Science" (Adams et al. 2015) that we guest edited for the *Journal of Social and Political Psychology*. Glenn Adams presented a preliminary version of the chapter at the "Decolonizing the Spirit" conference in April, 2015. The chapter benefitted greatly from discussions with colleagues from the Costa Rican Liberation Psychology Collective and the Cultural Psychology Research Group at the University of Kansas. Funding for the project came from a UCR-KU Collaboration Grant to Ignacio Dobles, Glenn Adams, and Ludwin E. Molina from the Vicerrectoría de la Investigación at the Universidad de Costa Rica and the Office of International Programs at the University of Kansas. Tuğçe Kurtiş also received funding in the form of a President's Development Grant at the University of West Georgia.

2. In part, this absence reflects the geopolitics of knowledge in hegemonic psychological science with an empirical basis in a narrow range of settings

(Arnett 2008; Henrich et al. 2010). However, this absence almost certainly also reflects *epistemologies of ignorance* (i.e., ways of knowing that promote not knowing of inconvenient truths; see Mills 2007) or more active processes of *unknowing* (i.e., convenient forgetting or silencing of empirical realities necessary to permit scientific business as usual; see Geissler 2013) about the centrality of racism and colonial violence in the constitution of psychological science and the Eurocentric modern global order.

3. Throughout this chapter, we make frequent use of "hegemonic psychological science" to refer to the institutions of scientific knowledge production associated with academic psychology. Research scientists within psychology departments popularized the use of *psychological science* both to distinguish themselves from practitioners of therapy and to wrap themselves in the legitimizing mantle of science. We use the term to refer to versions of the field, common in North American settings but increasingly dominant globally, in which ideologies of scientism are especially dominant (Denzin and Lincoln 2012, 2018). We add *hegemonic* to emphasize particular understandings and ways of being that originally derived from European (American) experience, but that colonial power has imposed as the default or natural standard for global humanity and mainstream academic work.

4. We follow others (Henrich et al. 2010) in the use of the acronym WEIRD to refer to the Western, educated, industrial, rich, and (supposedly) democratic settings that disproportionately inform intellectual imagination and knowledge production in hegemonic psychological science.

5. For present purposes, one can understand Orientalism as a form of epistemic (Spivak 1988) or epistemological violence (Teo 2010) by which intellectuals create a sense of cultural superiority and justify Eurocentric global domination by constructing an image of communities in the Global South as backward or deficient Others in need of (European) civilization (cf. Said 1978).

REFERENCES

Adams, G., & Salter, P. S. (2007). Health Psychology in African Settings: A Cultural-Psychological Analysis. *Journal of Health Psychology, 12,* 539–551.

Adams, G., Dobles, I., Gómez Ordoñez, L., Kurtiş, T., & Molina, L. E. (2015). Decolonizing Psychological Science: Introduction to the Special Thematic Session. *Journal of Social and Political Psychology, 3,* 213–238. https://doi.org/10.5964/jspp.v3i1.564.

Adams, G., Estrada-Villalta, S., & Gómez Ordoñez, L. (2018). The Modernity/Coloniality of Being: Hegemonic Psychology as Intercultural Relations. *International Journal of Intercultural Relations: Special Issue on Colonial Past and Intergroup Relations, 62,* 13–22.

Arnett, J. J. (2008). The Neglected 95%: Why American Psychology Needs to Become Less American. *American Psychologist*, *63*, 602–614.

Barrero, E. (2012). *Del Discurso Encantador a la Práxis Liberadora: Psicología de la Liberación. Aportes para la Construcción de una Psicología desde el Sur* [From Enchanting Speech to Liberatory Praxis: Liberation Psychology: Contributions toward the Construction of a Psychology from the South]. Bogotá, Colombia: Ediciones Catedra Libre.

Becker, D., & Marecek, J. (2008). Dreaming the American Dream: Individualism and Positive Psychology. *Social and Personality Psychology Compass*, *2*, 1767–1780.

Bulhan, H. A. (1985). *Frantz Fanon and the Psychology of Oppression*. New York, NY: Plenum.

Bulhan, H. A. (2015). Stages of Colonialism in Africa: From Occupation of Land to Occupation of Being. *Journal of Social and Political Psychology: Special Thematic Section on Decolonizing Psychological Science*, *3*, 239–256.

Burton, M., & Gómez Ordóñez, L. (2015). Liberation Psychology: Another Kind of Critical Psychology. In I. Parker (Ed.), *Handbook of Critical Psychology* (pp. 248–255). New York, NY: Routledge.

Burton, M., & Kagan, C. (2005). Liberation Psychology: Learning from Latin America. *Journal of Community and Applied Social Psychology*, *15*, 63–78.

Cole, T. (2012). The White-Savior Industrial Complex. *The Atlantic Monthly* (online). March 21, 2012. http://www.theatlantic.com/international/archive/2012/03/the-white-savior-industrial-complex/254843/.

Comaroff, J., & Comaroff, J. L. (1991). *Of Revelation and Revolution, Volume 1. Christianity, Colonialism, and Consciousness in South Africa*. Chicago, IL: University of Chicago Press.

Comaroff, J., & Comaroff, J. L. (2012). Theory from the South: Or, How Euro-America Is Evolving Toward Africa. *Anthropological Forum: A Journal of Social Anthropology and Comparative Sociology*, *22*, 113–131.

Comas-Díaz, L., Lykes, M. B., & Alarcón, R. D. (1998). Ethnic Conflict and the Psychology of Liberation in Guatemala, Perú, and Puerto Rico. *American Psychologist*, *53*, 778–792.

Connell, R. (2007). *Southern Theory: Social Science and the Global Dynamics of Knowledge*. Crow's Nest, NSW, Australia: Allen & Unwin.

Crenshaw, K., Gotanda, N., Peller, G., & Thomas, K. (Eds.). (1995). *Critical Race Theory: The Key Writings that Formed the Movement*. New York, NY: The New Press.

David, E. J. R., & Okazaki, S. (2006). Colonial Mentality: A Review and Recommendation for Filipino American Psychology. *Cultural Diversity and Ethnic Minority Psychology*, *12*, 1–16.

Denzin, N. K., & Lincoln, Y. S. (2012). Introduction: The Discipline and Practice of Qualitative Research. In N. K. Denzin & Y. S. Lincoln (Eds.), *The Sage Handbook of Qualitative Research* (4th ed., pp. 1–19). New York, NY: Sage.

Denzin, N. K., & Lincoln, Y. S. (2018). Introduction: The Discipline and Practice of Qualitative Research. In N. K. Denzin & Y. S. Lincoln (Eds.), *The Sage Handbook of Qualitative Research* (5th ed., pp. 1–26). New York, NY: Sage.

Dobles, I., Baltodano, S., & Zúñiga, V. L. (Eds.). (2007). *Psicología de la Liberación en el Contexto de la Globalización Neoliberal: Acciones, reflexiones y desafíos* [Liberation Psychology in the Context of Neoliberal Globalization: Actions, Thoughts, and Challenges]. San José, Costa Rica: Editorial Universidad de Costa Rica.

Dudgeon, P., & Walker, R. (2015). Decolonizing Australian Psychology: Discourses, Strategies, and Practice. *Journal of Social and Political Psychology: Special Thematic Section on Decolonizing Psychological Science, 3,* 276–297.

Duran, E., & Duran, B. (1995). *Native American Postcolonial Psychology.* Albany, NY: State University of New York Press.

Dussel, E. (1972). *Teología de la Liberación y Ética. Caminos de Liberación Latinoamericana* [Liberation Theology and Ethics: Roads of Latin American Liberation]. Buenos Aires, Argentina: Latinoamérica Libros.

Enriquez, V. G. (1992). *From Colonial to Liberation Psychology: The Philippine Experience.* Honolulu, HI: University of Hawaii Press.

Escobar, A. (2007). Worlds and Knowledges Otherwise: The Latin American Modernity/Coloniality Research Program. *Cultural Studies, 21,* 179–210.

Fals Borda, O. (1987). The Application of Participatory Action-Research in Latin America. *International Sociology, 2,* 329–347.

Fanon, F. (1963). *The Wretched of the Earth* (C. Farrington, Trans.). New York, NY: Grove Press. (Original work published 1961).

Fanon, F. (1967). *Black Skin, White Masks.* (C. L. Markmann, Trans.). New York, NY, USA: Pluto Press. (Original work published 1952).

Freire, P. (1993). *Pedagogy of the Oppressed.* New York, NY: Continuum. (Original work published 1970).

Geissler, P. W. (2013). Public Secrets in Public Health: Knowing Not to Know While Making Scientific Knowledge. *American Ethnologist, 40,* 13–34.

Grosfoguel, R. (2016). From 'Economic Extractivism' to 'Epistemical Extractivism' and 'Ontological Extractivism': A Destructive Way to Know, Be and Behave in the World. *Tabula Rasa, 24,* 123–143.

Gutiérrez, G. (1971). *Teología de la liberación. Perspectivas* [Liberation Theology: Perspectives]. Salamanca, Spain: Ediciones Sígueme.

Henrich, J., Heine, S. J., & Norenzayan, A. (2010). The Weirdest People in the World? *Behavior and Brain Sciences, 33,* 61–83.

Kağıtçıbaşı, C. (1995). Is Psychology Relevant to Global Human Development Issues? *American Psychologist, 50,* 293–300.

Kurtiş, T., & Adams, G. (2015). Decolonizing Liberation: Toward a Transnational Feminist Psychology. *Journal of Social and Political Psychology: Special Thematic Section on Decolonizing Psychological Science, 3,* 388–413.

Lacerda, F., & Dobles, I. (2015). La Psicología de la Liberación 25 Años después de Martín-Baró: Memoria y Desafíos Actuales. [Liberation Psychology 25 Years after Martín-Baró: Memory and Current Challenges.] *Teoría Y Crítica de La Psicología, 6,* 1–5. Retrieved from http://www. teocripsi.com/ojs/index.php/TCP/article/view/26.

Lander, E. (Ed.). (2000). *La Colonialidad del Saber: Eurocentrismo y Ciencias Sociales. Perspectivas Latinoamericanas* [The Coloniality of Knowledge. Eurocentrism and Social Sciences. Latin American Perspectives]. Buenos Aires, Argentina: Consejo Latinoamericano de Ciencias Sociales. Retrieved from http://bibliotecavirtual.clacso.org.ar/clacso/sur-sur/20100708034410/lander.pdf.

Lykes, M. B., & Moane, G. (2009). Editors' Introduction: Whither Feminist Liberation Psychology? Critical Explorations of Feminist and Liberation Psychologies for a Globalizing World". *Feminism & Psychology, 19,* 283–297.

Maldonado-Torres, N. (2007). On the Coloniality of Being: Contributions to the Development of a Concept. *Cultural Studies, 21,* 240–270.

Markus, H. R., & Kitayama, S. (1991). Culture and Self: Implications for Cognition, Emotion, and Motivation. *Psychological Review, 98,* 224–253.

Martín-Baró, I. (1986). Hacia una Psicología de la Liberación. [Toward a Liberation Psychology.] *Boletín de Psicología de El Salvador, 22,* 219–231.

Memmi, A. (1965). *The Colonizer and the Colonized.* New York, NY: Orion Press.

Mignolo, W. D. (2011). *The Dark Side of Western Modernity: Global Futures, Decolonial Options.* Durham, NC: Duke University Press.

Mills, C. W. (2007). White Ignorance. In S. Sullivan & N. Tuana (Eds.), *Race and Epistemologies of Ignorance* (pp. 13–38). Albany, NY: SUNY Press.

Montero, M. (2007). The Political Psychology of Liberation: From Politics to Ethics and Back. *Political Psychology, 28,* 517–533.

Montero, M., & Sonn, C. S. (Eds.). (2009). *Psychology of Liberation: Theory and Applications.* New York, NY: Springer.

Ngũgĩ wa Thiong'o. (1986). *Decolonising the Mind: The Politics of Language in African Literature.* Nairobi, Kenya: East African Educational Publishers.

Okazaki, S., David, E. J. R., & Abelmann, N. (2008). Colonialism and Psychology of Culture. *Social and Personality Psychology Compass, 1,* 90–106.

Phillips, N. L., Adams, G., & Salter, P. S. (2015). Beyond Adaptation: Decolonizing Approaches to Coping with Oppression. *Journal of Social and Political Psychology: Special Thematic Section on Decolonizing Psychological Science, 3,* 365–387.

Said, E. W. (1978). *Orientalism.* New York, NY: Vintage.

Santos, B. S. (2014). *Epistemologies of the South: Justice Against Epistemicide.* Boulder, CO: Paradigm Publishers.

Segalo, P., Manoff, E., & Fine, M. (2015). Working with Embroideries and Counter-Maps: Engaging Memory and Imagination Within Decolonizing

Frameworks. *Journal of Social and Political Psychology: Special Thematic Section on Decolonizing Psychological Science, 3,* 342–364.

Shweder, R. A. (1990). Cultural Psychology: What Is It? In J. W. Stigler, R. A. Shweder, & G. Herdt (Eds.), *Cultural Psychology: Essays on Comparative Human Development* (pp. 1–43). New York, NY: Cambridge University Press.

Smith, L. T. (1999). *Decolonizing Methodologies: Research and Indigenous Peoples.* London, UK: Zed Books.

Spivak, G. C. (1988). Can the Subaltern Speak? In C. Nelson & L. Grossberg (Eds.), *Marxism and the Interpretation of Culture* (pp. 271–313). Basingstoke, UK: Macmillan Education.

Teo, T. (2010). What Is Epistemological Violence in the Empirical Social Sciences? *Social and Personality Psychology Compass, 4,* 295–303.

Tomlinson, B., & Lipsitz, G. (2013). American Studies as Accompaniment. *American Quarterly, 65,* 1–30.

Tuck, E., & McKenzie, M. (2014). *Place in Research: Theory, Methodology, and Methods.* New York, NY: Routledge.

Watkins, M. (2015). Psychosocial Accompaniment. *Journal of Social and Political Psychology: Special Thematic Section on Decolonizing Psychological Science, 3,* 324–341.

Watkins, M., & Shulman, H. (2008). *Toward Psychologies of Liberation.* Basingstoke, UK: Palgrave Macmillan.

PART III

Education

CHAPTER 4

Reviving the Spirit by Making the Case for Decolonial Curricula

Kimberly L. Todd and Valerie Robert

Abstract This chapter explores the need for decolonial curricula to disrupt the hegemonic and colonial narratives of state curricula. It examines how hegemonic Western knowledge systems were birthed by Cartesian separations, and how these have seeped into Western education, enabling schooling to become a tool for colonial violence. It underlines a need for anti-colonial, decolonial and Indigenous frameworks in curricula, to disrupt the state colonial narratives. This chapter subsequently presents examples of what decolonial curricula can look like. These new decolonial curricula are designed to work in conjunction with state curricula as building blocks in the average classroom and to resolve Cartesian separations. The hope embedded in such blueprints of decolonial curricula is to foster pervasive love, consent, reciprocity and intent throughout the learning process for students and teachers alike in a bid to honour the diversity inherent in humanity.

K. L. Todd (✉)
Ontario Institute for Studies in Education, University of Toronto, Toronto, ON, Canada
e-mail: k.todd@mail.utoronto.ca

V. Robert
Ontario College of Teachers (OCT), Toronto, ON, Canada

© The Author(s) 2018
N. N. Wane and K. L. Todd (eds.), *Decolonial Pedagogy*,
https://doi.org/10.1007/978-3-030-01539-8_4

57

Keywords Curricula · Decolonial · Cartesian separation · Education

INTRODUCTION

The chapter explores the need for alternative decolonial curricula to disrupt the hegemonic and colonial narratives of state curricula in schools. The roots of this chapter are in the authors' lived experiences, which have forced them to confront forces of oppression and colonization in schooling. By first examining those experiences, it will build on them and explore hegemonic Western knowledge systems and how they have resulted in the oppressive nature of the schooling system in both the curricula and the school system itself, taking a toll mentally, emotionally and spiritually on both teachers and students alike. With this unveiling of the colonial logics comes a need to disrupt the Western epistemic norms that have silenced and devalued other knowledges that are intrinsically connected to human development and wellbeing. Here we posit the need for decolonial curricula to run in tandem with the state curricula as a means in which to rupture epistemic barriers by providing decolonial tools that initiates a pathway towards the embracing of other forms of knowledges that have been subjugated under the colonial logics. We explore the concept of decolonial curricula, what it can achieve, and how we imagine it working for educators in the classroom and beyond.

LOCATING OURSELVES

Kimberly

I am both a settler in a place of privilege on Turtle Island, and a Goan and Anglo Indian whose birth and identity is a direct result of colonization. I am an elementary educator who was drawn to the teaching profession because I enjoyed the learning process and working with students. My first experience in the classroom took place in South Korea where I taught English. Upon returning to Canada I gained acceptance into the B.Ed. program which proved to be eye opening. While doing my second practicum in a middle class, predominantly privileged and white neighbourhood, the school environment felt like the antithesis of who I was. Emotionally, physically, spiritually and mentally that short practicum experience felt violent in some way. At the time I grappled with the experience, my belonging and my place in education. I was viscerally confronted with the knowing that this is where diversity goes die, although it took me years to come to this understanding. Entering into the M.Ed. was rather serendipitous and it propelled me along a new path. It was in my M.Ed. program where I began to learn about anti-colonial and decolonial frameworks and the inherent oppressive structures that reside within the education system. For the first time in my life I was able to peek behind the curtain of colonial logics, and see the myriad ways in which colonialism had permeated all areas of my being. During my M.Ed. I met a friend, a like minded soul who had both a love of learning, an insatiable sense of curiosity and an unending urge to travel. Her name was Valerie and it became clear that we were both going to teach abroad. So together we embarked on a two-year contract.

Valerie

I am a settler living on unceded Algonquin territory in Ottawa, a Franco-Ontarian on both sides of my family who grew up white, cis-gendered, and in a middle-class family. By profession, I am a certified elementary educator. I initially chose to enter the teaching profession for my lifelong love of learning and participating in knowledge sharing. I was coming out of a degree in international development and was thinking of combining both fields of knowledge in a future career. My first teaching experience was pre-certification, as a volunteer at a summer school program in Cameroon. It is there that I was forced to start confronting some uncomfortable truths about my privilege and the underlying power dynamics I was participating in. These new realizations of my privilege stood in contrast with the narratives of victimization from my francophone upbringing and Canadian settler identity, where we posited ourselves as victims of oppression and often thought of ourselves as survivors of a culture war. I did not, however, have the language to express these new thoughts or engage fully with them yet. I left with a sense of disquiet and unease to start my B.Ed. My practicum, in a rural area, was an upsetting experience. I felt stifled, and on a deep level, distraught by the cultural sameness, the lack of diversity of thought and creativity. It felt mechanical, and from the first day, though I enjoyed teaching, I felt estranged. Again, I could not pinpoint the exact origin of those feelings, but I was finally forced to acknowledge out loud that I had serious doubts about what it was I wanted to do and my place in the world. This is what brought me to do an M.Ed. For the first time, I was exposed to an examination of structures of power, colonization, privilege and oppression that infiltrate and direct all aspects of our lives and society. In the last two courses of my M.Ed., I met Kim. Her energy and sense of adventure struck a chord with me, as did her thoughtfulness and curiosity about the world. We quickly realized that we were grappling with some of the same questions and shared a desire to step outside our comfort zone and expand our horizons towards the unknown.

OUR STORY

We set out eager to teach abroad, to live, learn and work in this new country. We began to feel the same sense of vague unease about teaching as we had felt elsewhere, and this feeling only grew more disconcerting with time. It became apparent that the same structures and systems of education that we experienced in Canada were found elsewhere. We were teaching students about frogs that they had never had the opportunity to see, the building structures were similar in the setup of the classroom, the curriculum was not designed by the local country and the school calendar followed the Western model, disregarding their different seasons and religion, and lastly the language we were teaching in was a colonial language. This happens across the globe because the modern model for schooling has been exported and transplanted from Western culture. It became obvious that one of the reasons Western curricula were being imported by other countries was to help their students go on to Western universities and be active participants in the world economy. The goal was never to create wholesome connected and healthy human beings. Schooling was transactional in nature and we wanted more for our students and for ourselves. This took a toll on us and we grappled with it at that time and we have grappled with it ever since. This has been deeply disconcerting for us, because this worldview is seemingly inescapable despite the wealth of diversity inherent in human experience. Schooling yet again proved to be inherently destructive to human diversity. Something needed to change…

The idea for decolonial curricula came to us one day while trying to create a lesson plan that dealt with teaching students about dreaming and decolonization. In the past connecting new themes to state curriculum had proved simple. Yet this time, when we opened the curriculum documents, it seemed impossible. We were going to have to change our entire lesson plan idea to fit the state curriculum. From that moment, we realized we needed alternative curricula to enable liberatory learning, one that could potentially run in conjunction with state curricula to disrupt the hegemony of Western knowledge.

THE HEGEMONY OF WESTERN KNOWLEDGE

Ramón Grosfoguel (2011) in his article "Decolonizing Post-Colonial Studies and Paradigms of Political-Economy: Transmodernity, Decolonial Thinking, and Global Coloniality" elucidates how Western Knowledge was constructed to be situated as the universal fount of knowledge. Grosfoguel (2011, p. 5) traces the origin point of this construct to around the time period when René Descartes pronounced "I think, therefore I am" and extended the role of God as the source of knowledge to the Western Man through the acquiring of scientific knowledge. Man was able to understand the mechanizations of nature and control nature in a way that prior to this period, remained solely in the realm of God (Grosfoguel 2011, p. 5). In the same article, Grosfoguel (2011) also notes that "by producing a dualism between mind and body and between mind and nature, Descartes was able to claim non-situated, universal, God-eyed view knowledge" (p. 5). As the Encyclopedia Britannica explains in its "Cartesianism" entry, the mind also encompasses the spirit, and the body encompasses matter (Watson 2016). This means that this schism also then extends to a separation between spirit and matter and a new means by which to interpret the world. The scientific paradigm born of this new mindset led to the ability to create new representations and models of the world. Mignolo (2011) in his book *The Darker Side of Western Modernity: Global Futures, Decolonial Options* provides the example of the *Theatrus Orbis Terarum* (p. 79). This was the first map created in which the human observer was placed above the Earth, at a distance, providing a God-eyed view (Grosfoguel 2011) in space which prior to this point was unimaginable (Mignolo 2011, p. 79). Space became conquerable in relation to the Earth, further perpetuating the notion of universality lending itself to the God-eyed view (Grosfoguel 2011; Mignolo 2011). Grosfoguel (2011) explains how the main attributes of a God-eyed view are the lack of speaker and the lack of location, leading to the creation of a false concept of universality and resulting in what is known as "...the "point zero" perspective of Eurocentric philosophies (Castro-Gómez 2003)" (p. 5). From the zero point of knowledge, Western Eurocentric knowledge is situated as universal, so any knowledge that possesses a speaker and or a location fails to be universal and is therefore considered inferior to Western Eurocentric knowledge, creating the hegemony of Western knowledge (Grosfoguel 2011; Mignolo 2011). As Mignolo (2011) writes, "The zero point is the site of observation from which the epistemic colonial differences and the epistemic imperial differences are mapped out" (p. 80). It is from this site of observation that humanity can be sorted out into human and sub-human.

The combination of the rise of science as the main epistemology and the creation of "point zero" (Grosfoguel 2011, p. 5) also granted new meanings to the concept of time as a means by which to degrade people who fall outside of the Western Eurocentric paradigm. As explained by Mignolo, cultures similar to Western European culture were considered to be closer to modernity, whereas cultures more dissimilar to Western European cultures were considered closer to tradition, hence placing them further back in time along an imaginary continuum towards progress and granting them an inferior standing by the nature of this remoteness (2011, p. 160). The idea of superior and inferior cultures utilizing the medium of space and time, in relation to Western European culture, was propagated by colonialism through violence (Mignolo 2011). This made colonialism the vehicle by which Western Eurocentric knowledge was spread throughout the world, through genocide, dispossession and assimilation, thus maintaining the hegemony of Western knowledges and devaluing, degrading and dehumanizing both the people whom were colonized and their knowledges in its wake.

HEGEMONIC AND OPPRESSIVE NATURE OF SCHOOLING

Through colonialism, education became one vehicle by which Western Eurocentric thought gained dominance and hegemony. One prime example of this was the residential school system in Canada. The document from the Truth and Reconciliation Commission of Canada, entitled *Canada, Aboriginal Peoples and Residential Schools: They Came for the Children*, says the following: "The Canadian government took on heavy responsibilities when it established residential schools. Education, it was said, would "civilize" Aboriginal people" (2012, p. 1). The repercussions of the residential schools are multigenerational and are still being felt and lived by today's First Nations, Inuit and Métis people. This is evident in the genocidal loss of lives, language, culture, tradition and wisdom (TRCC 2012). This document from the Truth and Reconciliation Commission of Canada says the following about the purposes behind the atrocities committed in the residential schools: "These were not side effects of a well-intentioned system: the purpose of the residential school system was to separate children from the influences of their parents and their community, so as to destroy their culture" (2012,

p. 1). Schooling in Canada was used as a colonial tool to enact genocide, violence, assimilation and cultural destruction. The colonial system built on Western Eurocentric thought was able to replicate and sustain itself by using and controlling the school system as a means by which to attempt to eradicate Indigenous culture, spirit and identity. The school system today, within a Canadian context and elsewhere, is born out of colonial logics and invariably perpetuates them. The modern school system hence wears the mantle of colonizer, espousing a single ontology and limiting the voices, modes and spaces of expression.

In this subsequent section we discuss the hegemonic and oppressive nature of the school systems in Canada and elsewhere by connecting the origins of how hegemonic Western knowledge shaped schooling. By schooling, we refer to a majority of formally organized and institutionalized education systems. We will trace how schooling was made in the image of Western knowledge systems by locating how the dualism of mind and body, by extension matter and spirit, mind and nature, as well as the Western Eurocentric conceptions of time and space are implemented within school systems.

One of the defining traits of school systems is the Cartesian separation of mind and body (Grosfoguel 2011, p. 5). The emphasis is on developing the mind separately from the body. This mentality is apparent in rules regulating students' movements. They must remain seated and focused, fidgeting is frowned upon and moving the body is generally perceived as a distraction to learning. For example, in an elementary and secondary context, only a short portion of the day is given to physical education and recess to take care of the body's natural drive to move and be active. The separation between mind and body is detrimental to the wellbeing of those whom it hopes to educate, due to being an unnatural, disjointed state of existence.

It is possible to follow the Cartesian separation even deeper. Within this separation, the mind is thought to encompass the spirit and the body to encompass matter (Watson 2016), hence with the separation of body and mind, so too follows the separation of spirit and matter. This has serious implications for the understanding of the world and its sacredness, which in turn also has direct consequences within the context of schooling. These consequences are most clearly demonstrated through the absence of spirituality. Spirituality is not valued, emphasized or practiced, unless it is a spirituality which exists within the confines of a specific religious context. With this exception, spirituality and the myriad ways in which humanity has engaged and currently engages spirit are excluded with negative repercussions that often go unseen. Njoki Wane (2007), in her chapter "Practicing African Spirituality: Insights from Zulu-Latifa, an African Woman Healer," writes, "This sense of our spirituality is where it all begins; it is our starting point, knowingly or unknowingly, it is the life force that informs our experience, our actions, our thoughts, and our very being" (pp. 48–49). When spirituality is erased from our lives it manifests as a sense of loss, as a hole that often remains unnameable, as a wound that cannot heal. This is especially true for our school systems for both students and educators as spirituality is an intrinsic part of life and learning.

The subsequent defining trait of the school system is the Cartesian separation of mind and nature. Within the school system the knowledge that is garnered, happens in isolation from nature. The majority of all learning happens within a building and in classrooms that for the most part are devoid of glimpses of the natural world. Learning takes places within walls, behind doors, with only windows to provide a vantage point from which to gaze upon the natural world, but not be part of it. When taking into consideration the gamut of schooling, very rarely is learning engaged outside the classroom. In some instances, aspects of the natural world are brought into classrooms, such as plants or classroom pets, but these are often placed within the periphery of the room, as part of the backdrop and not as a central focus. The very subjects that are taught, are taught in isolation from each other. In the 10th anniversary edition of David Orr's *Earth and Mind* (2004), in the introduction titled *What Is Education For?* he writes, "We have fragmented the world into bits and pieces called disciplines and subdisciplines, hermetically sealed from other such disciplines. As a result, after 12 or 16 or 20 years of education, most students graduate without any broad, integrated sense of the unity of things" (2004, p. 11). These separate subjects and disciplines only prove to further alienate the mind from nature by failing to demonstrate the interconnections and interdependence of our existence on this planet. This comes at a cost emotionally, spiritually, mentally and physically for us all. Therefore, the Cartesian separation of mind and nature is demonstrated with schooling in a variety of ways, subjecting educators and students to the hegemony of Western knowledge and its pervasive effects.

The next worrisome trait of the school system is time as a linear, colonial, and Western construct. The school day progresses in a rigid, clearly segmented and linear progression of time allotments. As Riyad A. Shahjahan (2014) writes in "Being 'Lazy' and Slowing Down: Toward Decolonizing Time, Our Body and Pedagogy" on the topic of time, "Consequently, with the introduction of the clock, time was delinked from human bodies, and human bodies from nature" (p. 490). With time at a premium and the emphasis being on packing in the curriculum, gone is the connection to our bodies' rhythms and the knowledge of our bodies "… as valid knowledge producers and elevated having its own value for generating focus, stillness, and more importantly, anchoring us in the 'now' moment" (Shahjahan 2014, p. 494). The short periods of time where the body is allowed to be unleashed are not considered to be learning time and are limited in length so as to maximize sit-down, class-room time. Furthermore, time in school systems is also wielded as an agent of colonial violence by voiding other cultural experiences of time. Linda Tuhiwai Smith (2012), in her book *Decolonizing Methodologies: Research and Indigenous Peoples*, remarks: "Representations of 'native life' as being devoid of work habits, and of native people being lazy, indolent, with low attention spans, is part of a colonial discourse that continues to this day" (p. 56). These colonial representations of time exist within the school systems as well, demarcating students as successful or unsuccessful.

The way in which space is utilized within school systems is also reflec-tive of a Western paradigm. Within schooling, space is heavily controlled. Many items within the spatial layout of a classroom are given property designations whether they belong to the school, teacher or students. Assigned seating is one example of this, wherein a false sense of personal property is given to the student that extends only so far. Space within the Western construct is perceived as being something to be conquered. Hierarchy is built into the classroom by the way in which teachers and students take up space within the classroom. Educators are placed at the forefront of the room and standing up, while desks are situated facing the educator and students are sitting down. The elevation of educator versus students demonstrates hierarchy. Moreover, the materials that take up the space are often synthetic, harsh, bleak and lacking artistic merit perhaps in an attempt to focus the mind. Elementary educators for example, often go to great lengths, spending their own money to try and enliven the room to improve the atmosphere for students, knowing this to be the case. It is also necessary to note that within a Canadian context, the land that schools reside on, whether they be elementary schools, high schools, or universities were obtained through colonial violence in an effort to dispossess Indigenous peoples of their land. Therefore, space within schooling is constructed along lines of control, property, hierarchy, dispossession and colonial violence, all of which are the progeny of Western knowledge systems that are inherently oppressive and exclusionary.

To expand on the theme of space, curricula in the school system occupies a space that is akin to the God-eyed view. Science and mathematics are given primacy over the arts and physical education, a fact most evident in the lack of funding of the arts within schools and its subsequent absence from school activities (Vargas 2017). These first two subjects were important in establishing the God-eyed view, and this makes them by default the perfect knowledge base to maintain the God-eyed view in today's education system, harkening back to Descartes' "I think, therefore I am" (Grosfoguel 2011; Mignolo 2011). Science and mathematics maintain this hierarchical position in part by never locating the origins of the learning standards put forth in curriculum documents. By omitting this information, they posit themselves as being universal. They also often exclude the body and the spirit as knowledge vehicles. These attributes effectively erase other knowledges and ways of being.

This section has sought to trace the origins of hegemonic Western knowledge and demonstrate its lineage into schooling. The full ramifications of the oppressive, exclusionary and hegemonic nature of schooling are vast and profound and are perhaps felt most strongly by educators and students. Patrick Chamoiseau (1997), in his book *School Days*, a fictionalized account of his schooling in Martinique, expresses some of those feelings by saying, "We went to school to shed bad manners: rowdy manners, nigger manners, Creole manners — all the same thing" (p. 120). Those who are caught in these systems, educators and students alike, try ardently to resist these impositions and to emerge from these systems as whole beings, hence the need for decolonial tools that can rupture the hegemony of Western knowledge.

RUPTURING EPISTEMIC BARRIERS TOWARDS A TRANSFORMATIVE PRAXIS OF EDUCATION

The above section has demonstrated how schooling has been born out of the oppressive and hegemonic nature of Western knowledge, therefore, moving forward we need decolonial tools to enable us to rupture these epistemic barriers and engage in liberatory praxis. Audre Lorde in *Sister Outsider* writes, *"For the master's tools will never dismantle the master's house"* (2007, p. 112). Her words speak a deep-seated truth, that we cannot continue building the foundations of our society, which begins within the school systems, with faulty building blocks. We need reimagined tools that are forged by different cosmovisions, epistemologies and ontologies that lie outside of the Western paradigm. We need tools that can begin to heal the rifts and wounds caused by the Cartesian separations and colonization. As M. Jacqui Alexander (2005) explains in her book *Pedagogies of Crossing*:

> Since colonization has produced fragmentation and dismemberment at both the material and psychic levels, the work of decolonization has to make room for the deep yearning for wholeness, often expressed as a yearning to belong, a yearning that is both material and existential, both psychic and physical, and which, when satisfied, can subvert and ultimately displace the pain of dismemberment. (p. 281)

To this end, we propose the creation and engagement of decolonial curricula born out of the hearts, minds, bodies and spirits that hegemonic Western knowledge systems have attempted to subjugate. This is our call to action.

DECOLONIAL DISCOURSES

It is important to continue to localize this call to action as being part of the decolonial discourse. Decolonial modes of resistance to colonial violence and oppression have been ongoing "...since the very inception of modern forms of colonization-that is, since at least the late fifteenth and early sixteenth centuries-..." (Maldonado-Torres 2011, p. 2). The twentieth century has been a time of profound decolonial uprising, restructuring and change, but it invariably remains "...an unfinished project..." (Maldonado-Torres 2011, p. 2). It is a century that has witnessed a new rash of decolonization in a number of regions and territories, such as the Caribbean and Africa, through the fall of formal colonial administrations, a series of events assisted by two World Wars, which dimmed Europe's power (Maldonado-Torres 2011, p. 2). The Bandung Conference took place in 1955, which spearheaded discourses on decolonization which would prove to be foundational (Mignolo 2011). Mignolo (2011) writes, "Decolonization departed from both capitalism and communism and opened up disobedient "third way" delinking from both. The aim of the conference was to promote economic and cultural cooperation and to oppose colonialism" (p. 59). Grosfoguel (2007) further expands on this by bringing to our attention that coloniality does not cease to exist because of the expulsion of formal colonial states/administrations, and that the path to decolonialized ways of being are far from over (p. 219). Decolonization is an unfolding and ongoing process that happens on multiple levels, both personally, structurally and systemically. Decolonial discourses have been ongoing and have led to new understandings and formulations of what the decolonial journey entails for individuals, communities and humanity. For example, *The Decolonial Turn* was coined at the 2005 conference in Berkeley California at the University of California (Maldonado-Torres 2011, pp. 5–6). Nelson Maldonado-Torres (2011) explains the core concept when writing, "... beyond the dialectics of identity and liberation, recognition and distribution, we have to add the imperative of epistemic decolonization, and in fact, of a consistent decolonization of human reality" (p. 4). This is where the necessity for decolonial curricula comes into play. We need epistemic decolonial tools to help us chart our course towards a reality steeped in the core vision of *The Decolonial Turn*.

What Can a Decolonial Curricula Achieve?

A call for decolonial curricula seeks to re-establish connections that were lost to Cartesian worldviews of separation and disconnection, creating a hegemonic worldview through schooling. Our vision is that decolonial curricula emerge as modes of resistance, renewal and resurgence to Western and colonial ways of being and living in the world; especially in regards to schooling. The truth is that decolonial curricula needs to arise from the local and from the speaker, that in no way attempts a bid for universality, rupturing the God-eye view (Grosfoguel 2011; Mignolo 2011). We imagine decolonial curricula sprouting up from knowledge keepers, elders, educators, scholars, artists, former students, anyone who has felt excluded, erased or unacknowledged by Western knowledge systems as decolonial curricula can be used in and out of the classroom. However, for the purpose of this chapter, we are primarily focusing on schooling systems.

School systems are generally organised around the false Cartesian premise that mind and body are separate entities. They have always been connected and interwoven, but Western hegemonic knowledge based on Cartesian dualism has taught us to ignore the interconnection of the mind and body and convinced us of their disconnect. A decolonial curricula honours the knowledge that mind and body are deeply entwined.

Likewise, matter and spirit have always been connected, but spiritual knowledges under a Western paradigm have been degraded. Spiritual knowledges provide access to the interconnected nature of the universe. Reviving spiritual knowledges such as dreaming, visioning, journeying, meditation, ceremonies and rituals to name but a few, provide mediums of communion with something greater than ourselves. Therefore, providing the space in which to revive spiritual knowledges within decolonial curricula opens up the heart and spirit to the interconnection of all life.

Mind and nature are also interwoven at a primal level, due to the mind originating entirely from nature. They can never be separated from each other, although Western hegemonic knowledge has forgotten this. Introducing curricula that espouses the connection between mind and nature moves us away from the anthropocentric viewpoint. An anthropocentric viewpoint is one in which the minds are disconnected from nature, leading to humans placing themselves at the center of the natural world. This creates the idea that we can control and own nature. When we develop a connection and relationship to nature and move away from anthropocentrism, we become more respectful and caring of the entire natural sphere, ourselves included.

Addressing time and space within schooling, we believe, requires systemic and structural change and re-envisioning of new decolonial schooling possibilities. Individual schools can work towards this goal, but decolonial curricula alone has its limits. However, the hope is that decolonial curricula can begin to make transformative shifts towards collective change so that schooling is revolutionized in ways that honour all our humanity and relationships.

How Does It Work?

The purpose of the decolonial curricula is to rupture the epistemic barriers of Western hegemonic knowledge. An educator, for example, looks at decolonial curricula first to provide concepts that fall outside of Western hegemonic knowledge, and then secondly refers to state curricula. As educators there is a responsibility to meet state curriculum objectives when working within state schooling. The decolonial model allows one to think outside the box by first centering decolonial values such as, but not limited to, for example, the connection between mind and body, or the connection between matter and spirit. A decolonial curriculum is supposed to be a tool to help liberate the soul, body, and mind by conceiving of new ways to educate. This tool can be used at any grade level and even outside of the school system, be it in corporations or workshops and etc.

A Guideline for What a Decolonial Curriculum Can Look Like

In this section we are going to outline a guideline for what one possible decolonial curriculum could potentially look like. The purpose is help educators envision their own decolonial curriculum. To begin, the decolonial curriculum should be broken down into main foundational concepts that can be based on the reconnection of the Cartesian separations of body from mind, spirit from matter, and nature from mind (Grosfoguel 2011; Watson 2016), but are not limited to these areas. These foundational concepts can act as an overall guiding principle and then be broken down into main objective or goals that have their own learning objectives. In the writing of decolonial curricula, models of objects in nature could be utilized to anchor the foundational concepts and to prevent a purely linear model. This could for example, be a visual that demonstrates the ecosystem of interrelationships and connections between the concepts, such as a seashell, a snowflake, a tree, and etc. This is important because invariably, when writing a decolonial curriculum, the concepts do not emerge in isolation and may flow into each other. We also suggest writing a decolonial curriculum in cooperation with the community, as the writing of it can be healing in and of itself. It is also necessary that anti-colonial frameworks be utilized, because our understanding is that anti-colonial frameworks unveil the colonial logics, whereas decolonial frameworks work towards healing, resurgence and renewal in the wake of ongoing colonial trauma. As to how to integrate an anti-colonial aspect, we leave that to reader, as each situation is unique. Below, we outline a specific example of what one potential foundational concept in a decolonial curriculum looks like.

One example could be reconnecting mind and body as a foundational concept. The main objective could be to have students turn their attention to the interplay between and mind and body. The decolonial learning objective can be to record their experience in a medium of their choice. Let us say elementary students have to learn about the digestive system, following state curriculum objectives. During lunch students could be asked to mindfully eat their food, taking in their senses in the body and their thoughts and emotions that arise from eating and also reflect on how their food landed on their plates via human labour. The recording part of the experience could be turned into assessment tasks that matches state curriculum objectives. In this way both objectives, those of the state curriculum and those of decolonial curriculum are met. They work together to provide students with an educational experience that acknowledges soul, mind and body. Additionally, it is important to note that Western knowledge is valuable in and of itself, but there needs to be a rebalancing of the power dynamics so that other knowledges are not devalued and excluded. There will be many difficulties in the writing of a decolonial curricula and it will be an ongoing process ripe with challenges, however, the potentialities for it to chart a pathway toward personal, communal and collective transformation are endless.

CONCLUSION

In conclusion, we have sought to chart the lineages between how Western knowledge became hegemonic and how this hegemony was embedded in schooling. We focused on the concepts around Cartesian separation that are inclusive of mind and body, matter and spirit, mind and nature, and time and space (Grosfoguel 2011; Mignolo 2011). We made the case for the need to reaffirm the connections that Cartesian dualism sought to separate and proposed the need for decolonial curricula to begin healing the rift that schooling is continually perpetuating. This has been a call to action for those who have suffered under the yoke of colonial oppression to reassert our own humanity. Decolonial curricula can potentially be a much-needed tool for educators, students, communities and beyond.

REFERENCES

Alexander, M. J. (2005). *Pedagogies of Crossing: Meditations on Feminism, Sexual Politics, Memory, and the Sacred*. Durham, NC: Duke University Press.
Chamoiseau, P. (1997). *School Days* (L. Coverdale, Trans.). Lincoln, NE: University of Nebraska Press (Original work published 1994).

Grosfoguel, R. (2007). The Epistemic Decolonial Turn: Beyond Political-Economy Paradigms. *Cultural Studies, 21*(2–3), 211–223.

Grosfoguel, R. (2011). Decolonizing Post-Colonial Studies and Paradigms of Political Economy: Transmodernity, Decolonial Thinking, and Global Coloniality. *Transmodernity: Journal of Peripheral Cultural Production of the Luso-Hispanic World, 1*(1).

Lorde, A. (2007). *Sister Outsider: Essays and Speeches*. Berkeley, CA: Crossing Press.

Maldonado-Torres, N. (2011). Thinking Through the Decolonial Turn: Post-Continental Interventions in Theory, Philosophy, and Critique—An Introduction. *Transmodernity: Journal of Peripheral Cultural Production of the Luso-Hispanic World, 1*(2), 1–15.

Mignolo, W. D. (2011). *The Darker Side of Western Modernity: Global Futures, Decolonial Options*. Durham, NC: Duke University Press.

Orr, D. W. (2004). *Earth in Mind: on Education, Environment, and the Human Prospect* (10th Anniversary ed.) (pp. xi–15). Washington: Island Press.

Shahjahan, R. A. (2014). Being 'Lazy' and Slowing Down: Toward Decolonizing Time, Our Body, and Pedagogy. *Educational Philosophy and Theory, 47*(5), 488–501. https://doi.org/10.1080/00131857.2014.880645.

Smith, L. T. (2012). *Decolonizing Methodologies: Research and Indigenous Peoples*. Dunedin: University of Otago Press.

Truth and Reconciliation Commission of Canada (TRCC). (2012). *Canada, Aboriginal Peoples and Residential Schools: They Came for the Children* (Catalogue No. IR4-4/2012E-PDF), pp. 1–54. Retrieved from http://www.myrobust.com/websites/trcinstitution/File/2039_T&R_eng_web[1].pdf.

Vargas, A. M. (2017). Arts Education Funding. *Journal of Women in Educational Leadership, 212*. https://doi.org/10.13014/K21Z42K0.

Wane, N. N. (2007). Practicing of African Spirituality: Insights from Zulu-Latifa, an African Woman Healer. In N. Massaquoi & N. Wane (Eds.), *Theorizing Empowerment: Black Canadian Feminist Thought* (pp. 47–54). Toronto: Inanna Publishers.

Watson, R. A. (2016). Cartesianism. In *Encyclopaedia Britannica Online*. Retrieved from https://www.britannica.com/topic/Cartesianism.

CHAPTER 5

Training for "Global Citizenship" but Local Irrelevance: The Case of an Upscale Nigerian Private Secondary School

Chizoba Imoka

Abstract Through the lens of a 16-year-old Nigerian secondary school student, this case study illustrates the human development and education crisis in Nigeria where the education system (represented by the private education sector) is focused on developing citizens for "global citizenship" but fails to prepare young Nigerians to learn about themselves as cultural-historical beings and agents for local and continental leadership. Although the school highlighted in this study is fictional, the experiences presented in the case are real. The case study was developed based on several research projects on private schools in Nigeria and my own personal schooling and advocacy experience in Nigeria. Focusing on the Magnum Prep Secondary School (MPSS) (pseudonym), the case highlights that the global citizenship curriculum measures its success based on the number of students who gain admission and attend Western universities. This pursuit is achieved by immersing the students in a colonial curriculum that upholds Western standards and values as the epitome of humanity. In effect, the curriculum denies these Nigerian students their

C. Imoka (✉)
University of Toronto, Toronto, ON, Canada

© The Author(s) 2018
N. N. Wane and K. L. Todd (eds.), *Decolonial Pedagogy*,
https://doi.org/10.1007/978-3-030-01539-8_5

humanity and right to equitably participate in the "global society" by not teaching them about their culture and history. Students' voice is also stifled in this school. For example, even though, students had critical and constructive views about their curriculum, student programming and student life, their views were not considered. In so doing, the school system fails to develop Nigerians who are academically nuanced and embody the capacity and human agency to drive human development in the country.

Keywords Curriculum · Global citizenship · Colonialism · Education

INTRODUCTION

Through the lens of a 16-year-old Nigerian secondary school student— Adaobi (Ada), this paper illustrates the human development and education crisis in Nigeria. Specifically, the paper explores how the high-cost private education system is focused on developing citizens for "global citizenship" but fails to prepare young Nigerians to learn about themselves as cultural-historical beings and agents for local and continental leadership. High-cost private schools are defined as schools where the school fees is at least CDN$3000/term. Although the school highlighted (Magnum Prep Secondary School [MPSS]), in this study is fictional, the experiences presented are real. In an effort to make readers have a deeper engagement with the long-standing impact of colonization on Nigeria's education system, the paper adopts a case study and storytelling approach. The research data that informs this paper is from three mixed methods research projects as well as my own personal schooling and advocacy experience in Nigeria.

Accordingly, the paper is divided into four main parts. "About Magnum Prep Secondary School" section provides a general overview of the school. "Adaobi: The Culturally Grounded Girl with Aspirations of Public Leadership" section introduces the 16-year-old protagonist— Ada. Ada comes from a very culturally grounded home and was disappointed to see how coloniality shaped the programming of her school. "Encountering the Colonial Character of MPSS and Disrupting the Myth of Excellence" section highlights how Ada mobilizes other students to speak up and engage the school for decolonial education changes. "Youth Mobilizing for School Reform" section focuses on the methodology and provides more details about data collection decisions. "Teaching Notes" section includes practical exercises for teachers to use in their reflective and teaching practice. The paper concludes with "Decolonial Teaching in Action" section.

About Magnum Prep Secondary School

MPSS is one of Nigeria's foremost and premium secondary schools. It is a grade 7–12 school located in the outskirts of Lagos state—Epe, Ogun State. Amongst the numerous achievements MPSS boasts of, it has produced numerous distinguished Nigerians ranging from computer innovators to politicians, economists, writers and public administrators. It is home to about 500 students. Majority of the students who attend the school are from relatively affluent backgrounds or/and have parents who have received western education and are very involved in their children's education.

Compared to the teachers in the low-cost private schools referenced in Tooley and Dixon's (2006), Tooley, Dixon and Olaniyan (2005) that are paid approximately $CDN20/month (NGN 6000), MPSS teachers are paid at least $500/month (NGN 140,000). The process of becoming an MPSS teacher is highly selective. MPSS teachers are recruited from various disciplines and universities. Potential MPSS teachers must be amongst the top 15% of their cohort. Before they commence teaching, they are retrained in MPSS's Teachers Training College.

Being ranked and perceived as one of the best private schools in Nigeria comes with heavy expectations from the parents and students. MPSS students are trained in at least three curriculums and sit for three intensive exams in Grade 12—IGSCE (Cambridge exams), American curriculum (SAT exams) and the West African/Nigerian Curriculum (WAEC and NECO exams).

In a recent interview on national television, the proprietress of MPSS—Mrs. Fresh (pseudonym) described her school as a "Hub for Nigerian global citizens". Our success, she argues is best captured by the number of MPSS graduates that are admitted into Canadian, British and American universities. With great pride, she commented on the number of A's her students produce in the British and American exams they write yearly. Upon graduation, at least 98% of all the students go abroad to study. To facilitate this process, a significant amount of resources is invested in teaching the students about life in the Western world.

For example, students get the opportunity to go for excursions to Abu Dhabi and leadership programmes in London. Summer school programs in Yale and Harvard are also heavily promoted. Students are required to participate in Western etiquette classes ranging from Small Talk, Dinner Networking and Speak Right, Sound Right—Queens English phonetics programmes. All grade 12 students are required to join the Duke of Edinburgh club in their school. MPSS's biggest and closet ally is a study abroad educational agency—The American Dream Institute (ADI). ADI sponsored MPSS's 2017 Inter-house sports, funds five prizes for academic excellence in grade 11/12 and waives the agency fee for study abroad applications for all MPSS students.

In the classroom, heavy emphasis is placed on Western scholarship. Students are taught about Shakespeare, Winston Churchill, British Parliament, American Government Policies, Western politics, etc. Concurrently, Nigerian academic content and language is downplayed or marginalized in the curriculum. Teachers are poorly trained to teach Nigerian content and it does not help that the Nigerian government scrapped History from the curriculum because Nigerian students allegedly do not have an interest in studying History (*Vanguard* 2014; Omolewa 2015). Students are not allowed to speak their indigenous language in school. In fact, some parents opt their children out of writing the Nigerian examinations because their children will be attending university abroad and, in their opinion "Western education best prepares their students to live and lead in the 21st century".

Because of the number of exams students have to write within a short time frame, culturally irrelevant Eurocentric curriculum content and MPSS's emphasis on exam centred student assessments, rote learning is strongly encouraged and relied on by the students.

Adaobi: The Culturally Grounded Girl with Aspirations of Public Leadership

Adaobi (Ada) had always had it in her to solve problems, to learn deeply, to ask question and to embrace her culture. She dreamt of being amongst the next generation of Nigerians who will bring pride to the country and continent whilst contributing to the advancement of humanity. Growing up, her nicknames were Odogwu Nwanyi (woman of substance, warrior woman), Enyi Nwa (Child of substance and Child of Courage) and Iron Lady; she is a strong young girl with courage and fire in her spirit. Family members tease that she is a spitting image of her grandmother who she was very fond of as a child. Ada's prowess in taking responsibility, mobilizing friends for solidarity and charitable social causes was all learnt from Mama's folktales and proverbs.

At home, Ada and her parents interact in her indigenous language (Igbo). Even though Ada's parents received their second degrees in England and the United States, they held their Igbo traditions very close to their heart and used it as a means to teach Ada at home. During the summer, Ada travelled with her parents to England to visit their aunties and cousins. Without fail, they visited their village at least three times in the year.

Amongst the many things Mama taught Ada, one thing that stayed with Ada was that "charity begins at home". Applying that philosophy to education meant that wanting to change the world had to begin with learning more about her country and hometown. Through the culture, traditions and values of the land, you will find the knowledge and wisdom to bring people together to solve community issues that emerge. Mama also shared stories about the invasion of White colonizers into Igboland and how it disrupted intricate life systems.

When it was time to go to senior secondary school, Ada was very excited because she was going to one of the best schools in Nigeria. She expected to grow academically, learn about Nigeria, strengthen her cultural understanding by building on what Mama taught her and through that lens, learn about the global society.

Encountering the Colonial Character of MPSS and Disrupting the Myth of Excellence

One of the first things that shocked Ada in her new school was the fact that students were not allowed to speak their indigenous language! It was forbidden to speak Igbo even with fellow Igbos or teachers in school. However, French was a mandatory subject for Grade 9 students but "indigenous languages" was optional (Nigeria Education Research Development Council, 2013). When she asked her teacher why she could not speak her language or learn more about her indigenous culture, the teacher's response was that: "we live in the 21st century and English is the language of the global world". Also, teachers had the habit of talking throughout the class period and leaving ten minutes out of a ninety minutes class for students to ask questions. Even though the school is aesthetically beautiful, Ada complained to her mum that it felt like a prison yard because students did not have any meaningful voice in the school. School hours were way too long (8–4 pm) and assignments were too much. The school did not protect or respect student's privacy. In addition, students were disciplined physically and emotionally when they did not get high grades. At the beginning of every term, the principal often announces the names of the top and "bottom" five students. The top students are praised and given an academic tie that distinguishes them from other students. The "bottom" students are humiliated either by flogging them or comparing them to other students. Ada had friends that were often part of the "bottom", they shared their pain and shame with Ada.

Contrary to the hand-on approach to leadership Ada is immersed in at home, the school prefects in MPSS were just figureheads. The available leadership positions were created by the school authorities and did not reflect the needs of the students or the school as a whole. For example, if students had a say in school governance, MPSS would have a Culture Prefect who would have lobbied for the infusion of Cultural Studies in the curriculum and the school as a whole. MPSS will also run a student radio to encourage students' voice and student input into the affairs of the school. If students had a say, MPSS will organize training for students to be more involved in community development and social entrepreneurship initiatives such as the TUC shop, the school farms and political organizing. The prefects did not have a say in the kind of classes they took, or the books in the library or the community service activities that students were required to do. In the hostels, the housemothers were often hostile and dismissive of student views.

Recently, Ada started questioning the "Academic Excellence" MPSS often boasts about. She observed that the school unofficially streams students by their perceived academic intelligence. All the "bright students" in a cohort were often streamed into the same class whilst the "not so bright students" were streamed to another class. Thinking out loud, Ada said to her friends, "if all the strong students are put in the same class and the weak ones in another class, how will students learn from each other? The students in the "not so bright" class will never know Math because no one else in the class does".

Ada complained to her parents about these issues, but they encouraged her to hang in there, as this is one of the best private schools in the country. Ada's patience was running out. She often wondered how she could live up to the popular Nigerian adage that "children are the leaders of tomorrow" with this quality of education that did not even let her practice leadership or teach her about her country or culture.

Thankfully, Ada found an African summer leadership program that stressed the importance of equipping African teenagers with the knowledge of African culture for Africa's transformation. For the entrance exam, she was required to place African countries on a map; Ada was stumped! She did not know her African countries. Ada realized how useless her MPSS education had been in teaching her about her continent's history and current affairs. In the excerpt below, Ada recounts that shameful experience:

> I know there is a Mozambique somewhere and you know, I think I put Kenya in an awkward place like where South Africa was supposed to go. It was really bad. I remember having a Malaysia there somehow. So, my knowledge about Africa was almost at a zero level. I was embarrassed and at the same time, I was like Wow, so this is how I grew up? (Imoka 2014, p. 97)

Youth Mobilizing for School Reform

Ada came back home furious. She knew that she had to speak up about this issue.

She reached out to other students who had graduated from MPSS to find out how useful their MPSS education had been to their university studies. MPSS alumni were equally bitter about their school experience and criticized how their school is considered one of the best in Nigeria. Ada decided to create a documentary, which she will share with current students of MPSS. Some students in the documentary said:

> At MPSS, you are taught to read and read and when it's exam time you pout out what you have read, give a few examples and that's it.......We have placed so much premium on the fact that "I went to MPSS" and the likes than the actual purpose of getting an education solid enough to be able to compete worldwide. (Adeyemi, Personal Communication, April 15, 2015)

In problematizing Nigeria's conceptualization of "quality schooling" and the school's claim to being premium, this student said:

> If you're going to be more expensive than other schools, you have to be more educative. It should be about how rich you are in actually educating your students and not because you look all flashy. If a school is going to be that expensive, I'm talking millions of Naira, it better be able to compete internationally with my mates in other schools. (Blessing, Personal Communication, April 20, 2015)

Clearly, MPSS's claim to quality education even for global citizenship is faulty because these students struggle in Western universities:

> I left my high school and I couldn't apply anything I learnt in the university here in Canada. Why? Because it was all jokes! Yes! From comprehension to composition in year 12!!! Yet, It's one of the most expensive school in Nigeria. While my year 12 mates over here were writing research papers, I was writing comprehension!!!! For our assignments, we were taught to write on silly topics such as: *How did you spend your holiday?* Or *What did Bisi say to Yemi?* and why? Rubbish!!! Instead of them to teach us how to critically analyze data, and write scholarly papers. If they will not do that, they should at least lay a foundation! (Bukky, Personal Communication, April 15, 2015)

Another student fired directly at the hypocrisy of MPSS's claim to global citizenship as she recalled learning more about Africa and Nigeria in her university in Canada. She did not even know that there were 54 countries in Africa. She did not know that Margaret Thatcher whom we all grew up praising called Nelson Mandela a terrorist! She also only learnt about the Biafra war through older White men in Canada and Chimamanda's book—*Half of a Yellow Sun* that was produced as a movie in 2014.

With these testimonials, Ada was ready to set up a strike action in school. Over the holiday, she posted her videos on Facebook and got a discussion started online. She started a petition for the school to increase students' voice, incorporate more indigenous culture and re-evaluate their curriculum. The plan was to gather enough signatures and present it at a perfect opening in school.

As soon as she arrived at school, the school authorities struck her last nerve; they rehired a catering company that the students had complained bitterly about. This caterer was notorious for serving students poor quality food. Ada and her friends agreed that this was the perfect opening to start the strike against the school. Secretly, they started making posters demanding for students' voice and indigenous cultural knowledge in the school. By dinner time, all 500 students across the school refused to eat the food and threatened to go on a longer hunger strike until the catering company was fired. Overnight all the posters demanding for students' voice and cultural knowledge were pasted around school. A copy of the documentary Ada created was left on the principal's desk. In the night, the students also agreed to camp out on the field after morning assembly till further notice. Ada sent a Facebook message to her family friend who is the education reporter for one of the national television stations in Nigeria (Nigerian Television Network [NTN]). By the morning, the press had flooded MPSS. All the students marched into school compound with different placards demanding students' voice in the school. Ada and the students presented their case to the members of the press and asked that the Nigerian government to send officials to her school to provide reasons why indigenous knowledge is not in their curriculum amongst other things. They all refused to go to class until the school authorities and government officials meet with them to watch the documentary, discuss their concerns and make radical changes to the curriculum and school culture right away.

Teaching Notes

This case study brings to light a potpourri of issues, dilemmas and opportunities in the Nigerian school system, which ironically, is often portrayed as "post-colonial". In every sense of the word, the Nigerian education system, gleaned from Ada's experience is far from post-colonial. Ada's education was informed by a colonial and Eurocentric vision of schooling that thrived on autocratic relations and the erasure of students' voice/identity.

As a result of Nigeria's defunct education system (evident in the missing Nigerian school girls (Chibok) and subsequent spade of insecurity crisis amid alarming rates of corruption), poor social infrastructure within the country and an elite centred social system, many high-cost private schools in Nigeria uphold a colonial curriculum that positions students for entry into Western higher education. This pursuit often comes at the expense of students' learning about their cultural heritage and history from an African centred perspective. For example, in Imoka (2016), I write about the student engagement experiences of three Nigerian secondary school graduates. Even though one of them lived in Lagos for 16 years, attended primary and secondary school in Lagos, Nigeria, she was never taught the history of Lagos state. However, she learnt about American and British history! Indigenous culture and history of Nigerians have been effectively divorced from their daily economic and social pursuits. Engaging with their culture is something left for their trips to the "village" during Christmas or experienced through food, dance and clothes. Even these cultural experiences have been coopted by non-African conglomerates. For example, popular "African" fabric—Hollandis is under the control of the Dutch company—Vlisco (Johnson 2013). Akinwumi (2008) in The "African Print" Hoax accuse Vlisco of deceiving consumers by coopting the African identity and attaching it to European produced material with origins in China, India amongst others. Vlisco's 2013 turnaround was USD378.4 million (How We Made it in Africa 2014, para 6). Effectively, Nigerian cultures and knowledge about the historical context are kept in a museum of sorts. Whilst in other countries, culture and historical knowledge is used to inform politics, economics and nurture citizen's capacity to engage in constructive politics, the opposite is true. The result of this is a human development crisis where a critical mass of Nigerians are going to school but are learning so little about their local context and thus incapable of defining their identity and being effective citizens within their home countries.

The case study draws from three research projects that I conducted between 2013 and 2017 on student experiences in the African schooling context. The experiences that are foregrounded in this paper are from three research participants that I interviewed. One of the participants regularly shared her schooling experiences on Facebook, I gathered data from her Facebook conversation. All the participants provided permission for their stories to be included in this paper. My interest in this area of study is personal. I completed my primary and secondary school education in Nigeria. The curriculum I was exposed to was Eurocentric and Nigerian indigenous cultures were never engaged as a critical site of information. In school, I observed that students' voice was stifled, and transformative leadership opportunities were limited for students. Somehow, I found the courage to push back against oppressive school rules and fought for change in school. The victories I won in school and the confidence I grew in was nurtured by a few teachers that did not punish me for being outspoken. The decision to present this data via case study method and storytelling is inspired by that experience.

Lynette Shultz's paper—Educating for Global Citizenship: Conflicting Agendas and Understandings also shaped my thinking. In this paper, she lucidly presents three conceptions of global citizenship (Neo-liberal, Radical and Transformationalist), outlines their ideological underpinnings and policy examples. In reading the paper, I realized that my education had prepared me to become a neo-liberal global citizen, but I wanted to become a transformationalist global citizen. The personal and theoretical dimensions of the paper reinforced the importance of presenting critical social justice issues in a way that allows readers to make meaningful and personal connections. I believe the storytelling approach gives readers an opportunity to see behind the theory and feel the complex ways in which colonial visions of schooling plays out in students' lives. Also, teachers have a lot of power, they can direct a student towards maintaining status quo or disrupting status quo. The storytelling approach is designed to draw teachers in and the exercises below will give teachers an opportunity to reflect on their practice from a personal and professional standpoint.

Decolonial Teaching in Action

This section of the paper includes exercises that teachers can use to reflect on their practice and orient students towards decolonial approaches to social justice engagement.

Exercise 1: Championing Critical Community Action—Justice Not Charity!

Ada's case has implications for the Canadian schooling context and curriculum especially as it relates to international development projects students are invited to participate in. Youth-focused international development programs organized by institutions like Free the Children often orient students towards a simplistic and paternalistic analysis of social problems in Africa. Poverty, for example, is presented as an individualized problem that can be wholly dealt with at an individual level through goat sponsorship or borehole creation. The problem of education in Africa is mischaracterized as an access problem, so the proposed solutions include mobilizing volunteers to build more schools and to raise funds to send more people to school. This perspective is problematic on multiple levels. First, it negates the structural and historic issues of colonialism that Ada's story clearly illustrates. Ada's story also illustrates the numerous ways colonial education negatively influences the identity formation of Nigerian youths and hampers their ability to assume culturally relevant leadership positions within their communities. Secondly, it sidelines grassroots movements in Africa such as the one Ada started to mobilize students and engage school authorities for change. At other levels of African society, there are numerous movements pushing for structural and decolonial social change. For teachers to orient students towards social justice, students must be empowered to develop a critical and historic perspective when thinking of international social issues. Doing this will position students and teachers to support grassroots decolonial struggles aimed at bringing about justice, promoting critical inclusion for all and relevant education for African youth.

The complicity of western countries in colonial education in Nigeria was made evident in how Ada's secondary school curriculum prepares her for Eurocentric exams that facilitates entry into Western universities. This interconnection shows the global nature of coloniality and colonial education. Even in Canada, colonizing and Eurocentric education is mainstreamed. This is especially the case amongst indigenous communities that go through the mainstream education system not ever learning about their histories or culture. When it is taught, it is taught in a Eurocentric way.

Teachers need to help students see the global nature of social problems and encourage students to take critical action both locally and internationally. This may involve conducting research to learn about a particular social issue from an anti-colonial lens and engaging politicians with that knowledge.

This exercise invites teachers to reflect on their practice. The following can serve as guiding questions:

1. How are MPSS and the experiences shared above similar or different from my schooling context and teaching experience?
2. What are the factors that enable this orientation of global citizenship that promotes teaching for western citizenship and local irrelevance?

3. In involving my students in international development projects in Africa, how can I change my engagement approach so that we ally with African youths to transform their societies?
4. In involving my students in international development projects in African, how can I engage my students in a political advocacy project that calls for the Canadian international development dollars to be channelled towards African centred education reform?

For question 2, teachers may read Lynette Shultz's (2007) Educating for Global Citizenship: Conflicting Agendas and Understandings (see reference list). For question 4, Julie Cammarota and Michelle Fine's book (2010)—*Revolutionizing Education*—Youth Participatory Research will be helpful (see reference list).

Exercise 2: Mentoring for Transformational Youth Action

Ada's orientation towards negotiating the shortcomings of her school was in many ways radical and outside of the norm. Instead of aligning with the oppressive school rules or allowing herself to be pushed out of the system, she asserted her agency and resisted the system in numerous ways. Ada's resistance is reminiscent of Cammarota and Fine's (2010) definition of Transformational Resistance. The authors explain below:

> A transformational approach to resistance moves the student to a deeper level of understanding and a social justice orientation. Those engaged in transformational resistance address problems of systematic injustice and seek actions that foster the greatest possibility for social change. (p. 4)

Youth Participatory Action Research (YPAR) has been identified by Cammarota and Fine (2010) as an effective research methodology for youth to learn about how injustice in society is produced and how social change can be achieved by youth. YPAR is often carried out with adult allies who serve as mentors to the youth. This exercise invites teachers to support and mentor students towards transformative projects. Teachers will guide students to identify social issues within their schools/ community at large, research the issue and then take action on it. In this engagement with students, the end goals are:

1. Help students conduct a personal and communal reflection on their schooling experiences and school climate in general.
2. Help students develop a needs list and needs assessment from their reflection process.
3. Support students to plan a workshop to strategize on how they could solve these problems.
4. Support students to take action on the identified problems.

In doing this, teachers should draw on their experience, read Chapter 9 of *Revolutionizing Education—Youth Participatory Action Research* (2010) and a chapter of choice from *Everyday Antiracism: Getting Real About Race in School* (2008) by Mica Pollock. McMahon and Portelli's (2004) article—*Engagement for What? Beyond Popular Discourses of Student Engagement* is an instructive read for conceptualizing different approaches and aims of student engagement.

Exercise 3: Teachers as Decolonial Public Intellectuals

As teachers, you may be presented with a request/opportunity to reform a school like MPSS. This could occur at a policy and school level. What will you do? How may you go about thinking through the problem? What does an alternative curriculum look like? What key considerations will you bring into the deliberation? Accordingly, the purpose of this exercise is twofold. First, to give teachers an opportunity to reflect on their role in facilitating decolonial curricula reform. Secondly, to empower teachers with preliminary knowledge and the language to advance decolonial education conversations. For this exercise, teachers should read Banks and Tucker (1998) *Multiculturalism's Five Dimension*'s article and Chapter 7 of Dei, Zine and James-Wilson (2002) *Removing the Margins*. For context, teachers should also read Chapters 1 and 2 of the *Nigeria's Philosophy of Education* (Nigeria Educational Research Development Council 1972). Below is a brief overview of the key concepts from the suggested readings.

In *Removing the Margins: The Challenges and Possibilities for Inclusive Schooling*, Dei et al. (2000) argue for an integrated schooling context that implicates students' language and identity in the process of their learning. He also outlines a conceptual framework for creating an inclusive school that represents silenced minorities and knowledge. This framework includes:

1. Visual representation of the cultures of students in the school's visual landscape. This also includes representation in books, posters and artwork amongst others.
2. Representation of non-European cultural knowledge within the school systems. Doing this validates the contributions of non-Europeans to the civilization of the world and enriches the knowledge base of the students.
3. Staff diversity enables students from diverse backgrounds and interests to identify role model teachers who they can build a relationship with while making their learning more meaningful (Dei et al. 2000, p. 176).

In *Multiculturalism's Five Dimensions*, Banks and Tucker (1998) identifies a whole school approach to enhancing multicultural education. At the core, he argues that the goal of developing global citizens does not preclude educating students about their local context. In fact, multicultural education done properly will help students become critical thinkers, intellectually curious and knowledge creators. These learning outcomes contribute significantly to the academic success of students. He outlines five dimensions in multicultural education (Banks and Tucker 1998). They include:

1. *Content integration*: The first step to multicultural education is integrating alternative knowledge (non-European, knowledge that represents the cultural heritage of students in the classroom) into the curriculum.
2. *Knowledge construction*: Students are challenged to engage the knowledge integrated into the classroom by helping them investigate the assumptions of the knowledge, the perspectives that are centred in the knowledge, the context in which it was created and the implications in their lives.
3. *Equity pedagogy*: To accommodate the learning styles and cultural diversity of students in the classroom, teachers should have dynamic teaching methods that enable different students to relate to the teaching material.
4. *Prejudice reduction*: Teachers have the responsibility to ensure that students show respect for other students who possess different knowledge or/and come from a different cultural background
5. *Empowering school culture and societal structure*: The diversity of the teaching staff and the school environment has to correspond

and align with the tenets of the multicultural education going on in the classrooms. Proactive steps should be taken to ensure that all students are able to participate equally in school activities and achievement is not exclusive to a group of students. A diverse teaching staff enables quality student interaction with teachers (Banks and Tucker 1998, pp. 1–7).

Drawing from the insights in these readings, teachers may do the following activities:

1. *Role play debate*: Simulate a talk show comprising of government officials, MPSS school owner and consultants and discuss the possibilities and benefits of MPSS changing their approach to student engagement and academic success.
2. *Practicalizing decolonial praxis*: To illustrate the practicality and advantages of engaging MPSS students in a culturally relevant way, teachers can create a lesson plan for a global citizenship class and outline the implications of such plan in the current MPSS arrangement.

The UNESCO's General History of Africa and Women in African History curriculum (provided in the reference list—UNESCO 2014/2014a/2014b) are practical examples of African centred material that could be used to develop culturally relevant lesson plans. These documents challenge mainstream perceptions on the availability of quality African centred historical materials about social life in precolonial Africa.

CONCLUSION

This chapter shows the boundless possibilities for hope and resurgence when students are socialized to be intellectually curious, intellectually honest and made to understand their responsibility to speak out and take action. Because Ada spoke out and engaged with other students, it inspired student communities to reflect on their experiences and mobilize for action. In so doing, it also pushed adults to action. Unfortunately, the education system in Nigeria and beyond is steeped in and framed by colonial understandings of education which seeks to control and reproduce colonial subjectivity. These subjectivities only serve to obliterate indigenous epistemologies/ways of knowing from global consciousness whilst maintaining a Eurocentric social order globally. Tragically, this warped understanding of education has been mainstreamed and consequently, "education" is uncritically accepted and engaged as a universal good. It's content, purpose and outcomes are not questioned.

In this light, encouraging students' voice is just as important as disrupting colonizing and oppressive conceptions of schooling and community service. Whilst the philosophical underpinnings of education as a human right is applicable globally, the purpose, content, orientation and outcomes of education is not. Education is political, historical and must be context specific. An education system and its corresponding components must be responsive to the history, aspirations and sociopolitical ideals of the social contexts in which it is delivered. In Canada, organizations like Free the Children provide a platform for Canadian students to act but not to change global colonizing institutional structures and oppressive social relations. Teachers and well-meaning individuals have a responsibility to understand how the current global education system serves to reinforce Eurocentric ways of knowing and oppressive social relations between the "custodians" of European knowledge and/ resources and citizens of the global south who are decentered from their cultural locations and sent off on a never-ending race of becoming a "global citizen". Teachers and well-meaning individuals/organizations must begin to think through and act out the local implications for social justice and community service at a global level. A good place to start is immersing students in participatory action research methodologies and preventing Canadian teenagers/organization from travelling abroad to "save" African children or to colonially educate the girl child! The problem is much more than access to colonizing education, it is a problem of understanding what it will take to create access to decolonizing education globally. Achieving this is a tall feat but one that is surmountable with an intergenerational global coalition for social justice.

References

Akinwumi, T. M. (2008). The "African Print" Hoax: Machine Produced Textiles Jeopardize African Print Authenticity. *Journal of Pan African Studies*, 2(5), 179–192.

Banks, J. A., & Tucker, M. (1998). Multiculturalism's Five Dimensions: Dr James A. Bank on Multicultural Education. *NEA Today*, 17(1), 1–7. Available at https://www.learner.org/workshops/socialstudies/pdf/session3/ 3.Multiculturalism.pdf.

Cammarota, J., & Fine, M. (2010). *Revolutionizing Education: Youth Participatory Action Research in Motion*. New York, NY: Routledge.

Dei, G. J. S., Zine, J., & James-Wilson, S. V. (2000). Representation in Education: Centring Silenced Voices, Bodies and Knowledges. In *Removing the Margins: The Challenges and Possibilities of Inclusive Schooling* (p. 176). Toronto: Canadian Scholars' Press.

Dei, G. J. S., Zine, J., & James-Wilson, S. V. (2002). *Inclusive Schooling: A Teacher's Companion to Removing the Margins*. Canadian Scholars' Press.

How We Made It in Africa. (2014). How Dutch-Based Vlisco Became One of Africa's Most Popular Fashion Companies [Blog]. *How We Made It in Africa*. Available at https://www.howwemadeitinafrica.com/how-dutch-based-vlisco-became-one-of-africas-most-popular-fashion-companies/43676/. Accessed December 22, 2017.

Imoka, C. (2014). *The Case for an African Centered Education System in Africa: A Case Study on African Leadership Academy, South Africa* (Masters' dissertation). University of Toronto, Canada.

Imoka, C. (2016). Student Engagement Experiences in Nigerian Secondary Schools: An Anti-Colonial and Student-Centered Analysis. In G. J. S. Dei & M. Lordan (Eds.), *Anti-Colonial and Decolonial Praxis* (Chapter 7). New York, NY: Peter Lang.

Johnson, R. (2013). Vlisco, the African Fashion Titan from Holland. *The Business of Fashion*. Available at https://www.businessoffashion.com/articles/global-currents/vlisco-the-african-fashion-titan-from-holland. Accessed December 22, 2017.

McMahon, B., & Portelli, J. P. (2004). Engagement for What? Beyond Popular Discourses of Student Engagement. *Leadership and Policy in Schools, 3*(1), 59–76.

Nigeria Educational Research Development Council—Adaralegbe, A. (1972). *A Philosophy for Nigerian Education: Proceedings of the Nigeria National Curriculum Conference, 8–12 September 1969*. Ibadan: Heinemann Educational Books (Nigeria).

Nigeria Educational Research Development Council—Federal Republic of Nigeria. (2013). *National Policy on Education—6th Edition*. Lagos, Nigeria: NERDC Press.

Omolewa, M. (2015). *Assault on the Teaching of History in Nigerian Schools*. *Guardian.ng*. Retrieved October 10, 2016, from http://guardian.ng/features/focus/assault-on-the-teaching-of-history-in-nigerian-schools/.

Pollock, M. (Ed.). (2008). *Everyday Antiracism: Getting Real About Race in School*. New York: New Press.

Shultz, L. (2007). Educating for Global Citizenship: Conflicting Agendas and Understandings. *Alberta Journal of Educational Research, 53*(3), 248–258. Available at http://myaccess.library.utoronto.ca/login?url=https://

search-proquest-com.myaccess.library.utoronto.ca/docview/228671594?
accountid=14771.

Tooley, J., & Dixon, P. (2006). 'De Facto' Privatisation of Education and the Poor: Implications of a Study from Sub-Saharan Africa and India. *Compare, A Journal of Comparative and International Education, 36*(4), 443–462. https://doi.org/10.1080/03057920601024891.

Tooley, J., Dixon, P., & Olaniyan, O. (2005). Private and Public Schooling in Low-Income Areas of Lagos State, Nigeria: A Census and Comparative Survey. *International Journal of Educational Research, 43*(3), 125–146.

Unesco.org. (2014a). *Pedagogical Use of the GHA | United Nations Educational, Scientific and Cultural Organization.* Retrieved December 28, 2014, from http://www.unesco.org/new/en/culturethemes/dialogue/general-history-of-africa/pedagogical-use-of-the-gha/.

Unesco.org. (2014b). *UNESCO | Women in African History.* Retrieved December 28, 2014, from http://en.unesco.org/womeninafrica/.

UNESCO. (2014). *Women in African History.* Retrieved from https://en.unesco.org/womeninafrica/. Accessed February 14, 2018.

Vanguard. (2014, March 12). *History Ends in Nigeria.* Available at http://www.vanguardngr.com/2014/03/history-ends-nigeria/.

CHAPTER 6

Using Arts-Based Learning as a Site of Critical Resistance

Marilyn Oladimeji

Abstract People who are illiterate are unable to participate in formal learning because they cannot understand the written form of content or communication. As a result, creative confidence is not nurtured or appreciated. As well, people in communities are unable to link formal learning or the lack of it to any aspect of their lives. This is more pronounced in areas where the occupation of a family is closely connected to cultural heritage and traditional knowledge. Ultimately, a lot of people drop out of school because formal education in schools has no connection whatsoever to their ways of living. Arts-based learning, with a focus on change rather than production, can bridge the gap between learning and life; for example, storytelling can be used as a medium for studying government policy. Because of the interruption among people's cultural heritage, life orientation, and the curricula, local and Indigenous knowledge that is culturally entrenched in the social fabric of communities is being eroded. Moving away from traditional know-how and cultural knowledge has thrown many communities into obscurity and peril. It is, obviously, very important for the local community in which the bearers of such knowledge live and produce. Systems of formal education need to recognize,

M. Oladimeji (✉)
The National Guild of Hypnotists, Toronto, ON, Canada

© The Author(s) 2018
N. N. Wane and K. L. Todd (eds.), *Decolonial Pedagogy*,
https://doi.org/10.1007/978-3-030-01539-8_6

value, and appreciate traditional and cultural knowledge in their interaction with local communities. Traditional knowledge also constitutes part of global knowledge.

Keywords Art-based learning · Storytelling · Indigenous knowledge · Learning culture · Decolonization

INTRODUCTION

In this chapter, I will discuss how using art-based learning creates opportunities for critical resistance. Art practice navigates the terrains of cultural translation and draws on the experiences of living cultures. It investigates the processes involved in cultural transfer and transformation by exploring theoretical frameworks. When I began to think about this chapter, I focused on the value placed on institutional learning versus Indigenous knowledge. The context in which I would like to share this information is based on the importance of considering Indigenous knowledge as a valuable pedagogy.

What is decolonization? As with anti-racism, decolonization is popularly misunderstood as a simple process involving the return of or recompense for historically stolen possessions, ideas, or resources. As with anti-racism, this misunderstanding of the aims and process of decolonization often leads to a characterization of anti-colonialism that situates it as reactionary, exclusionary, violent, and uncomplicated (Murad 2010).

Anti-colonialism and decolonization, however, are far more complicated processes, addressing much more than the physical spaces and resources plundered during ongoing colonial and neo-colonial ventures (Said 1994). As many anti-colonial theorists have argued, the process of colonization is always connected to—though not limited to—the material (Ahmed 2000; Said 1994; Sardar 1999; Meyer et al. 1993). Colonial processes have also targeted culture, knowledge, understanding, and all of their processes of social production. Indigenous knowledge and non-Western knowledge have been uprooted, criminalized, exterminated, pilfered from, rendered obsolete, dehumanized, and very firmly relegated to the margins. As Said (1994) argues in "Culture and Imperialism," colonial processes have relied on and continue to rely on cultural imperialism in order to articulate and proliferate the ideology that one culture, one system of knowing, one way of living is superior to all others and, thus, is bound to save, convert, or exterminate its inferior competition. In so doing, cultural imperialism has also supplanted diverse discourses of hoping with dominating and colonizing others. Every element of how we gain knowledge and learn to exist within our knowledge is subject to colonizing agendas, and Indigenous communities and communities of color are different from and maintain resistance to this hegemony at a great cost to themselves and their people. It is an anti-colonial struggle to seek to retain and actively engage in alternative thought processes and methods of reasoning and resistance. This effort is fraught with risk, and those who engage in it and who are most affected by cultural, social, and physical colonialism have a multitude of solid reasons not to share their endeavors with academics. As Smith (1999) notes, the act of studying Indigenous knowledge and cultural anti-decolonization is also the act of reifying, intellectualizing, translating, and neutralizing these efforts. Therefore, the work of decolonization is rarely an academic process. However, pushing these boundaries within academia in ways that do not compromise the integrity of Indigenous knowledge is important.

In *Looking White People in the Eye*, Razack (1998) takes on anti-oppression education as it functions in community, classroom, and active spaces—teasing out its many contradictions and limitations. She explores the pitfalls of the narrative focus that anti-oppression education brings to learning spaces, deconstructing the desirability of such narratives as part of the colonizing framework. Razack points out that which narratives are valued, who tells the story, and what role the narratives end up playing in the educational space actually serve to reinforce dominant paradigms. Starting with some definitions so that we are on the same page, knowledge is often seen as a rich form of information. A useful definition of knowledge is that it is about know-how and know-why. The recipe, though, would be: (1) written knowledge—explicit knowledge, relevant knowledge, and experience and (2) skill knowledge—knowledge that is in the head and is not easily written down. Many definitions of knowledge management exist. A common definition is: "The collection of processes that govern the creation, dissemination, and leveraging of knowledge to fulfill organizational and community purposes" (Gurteen 1999).

Indigenous knowledge is local knowledge: knowledge that is unique to a given culture or society. Indigenous knowledge contrasts with the international knowledge system generated by universities, research institutions, and private firms. It is the basis for local-level decision-making, health care, education, natural-resource management, and a host of other activities in communities (Warren 1991). Indigenous knowledge is the information base of a society and facilitates communication and decision-making. Indigenous information systems are dynamic and are continually influenced by internal creativity and experimentation, as well as by contact with external systems (Flavier 1995, p. 479).

Why is Indigenous knowledge important? The basic component of any country's knowledge system is its Indigenous knowledge. It encompasses the skills, experiences, and insights of people, applied to maintain or improve their livelihood (World Bank 2007).

Indigenous knowledge is developed and adapted continuously to gradually changing environments and is passed down from generation to generation and closely interwoven with people's cultural values. Indigenous knowledge is also the social capital of the poor; their main asset to invest in the struggle for survival, to produce food, to provide shelter, or to achieve control of their own lives.

Practices vanish as they become inappropriate for new challenges or because they adapt to slowly. However, many practices disappear only because of the intrusion of foreign technologies or development concepts that promise short-term gains or solutions to problems without being capable of sustaining them (Mabule 2016). Overcoming the fear of difference is an important aspect and so is developing a kind of personal security that is not easily threatened. This personal security comes from trust in oneself, trust in the other, and a larger trust in the ongoing process of Indigenous knowledge. The human story has always included music, movements, enactment and ceremony, and creating sacred objects and images in sand and stones. In traditional cultures, healers were musical artists and story tellers, and they enacted rituals and ceremonies for healing the community. The realm of the arts is much older and broader than that of the social sciences.

Paolo Knill et al. (2004) referred to our capacity to respond to what is beautiful as our aesthetic responsibility—an ethical call to care for the beauty of life and a fundamental aspect of what it means to be a human being. Changing a culture is tough. Not only does it mean change, which has always been difficult, but it also means seeing the world in a different way. It means revealing our hidden paradigms, such as the tacit acceptance that "knowledge is power." The real answer is to help people to see for themselves that knowledge sharing is in their personal interest.

Although the old paradigm was "knowledge is power," today it needs to be explicitly understood that "sharing knowledge is power." If people understand that sharing their knowledge helps them perform their jobs more effectively, helps them to retain their jobs, helps them in their personal development and career progression, rewards them for getting things done, and brings more personal recognition, then knowledge sharing will become a reality. We start by thinking like artists.

Years of oppression by colonizing powers, initially involving wide-scale conquest and slavery, afterwards consisting of forcible changes to Indigenous peoples' ways of life, and, most recently, characterized by the expropriation of Indigenous knowledge and bio-piracy of natural resources used and protected by Indigenous peoples, have put huge pressures on the relationship between Western and Indigenous societies (Dunn 2014). These difficulties are mirrored in the ways the two distinct knowledge systems view each other. Western knowledge systems are built upon the idea of positivism, which is the belief that the most trustworthy source of knowledge is information acquired through the senses and verified by logical, scientific, or mathematical testing (Dunn 2014). Knowledge that does not come in this way is regarded with a great deal of suspicion. Indigenous knowledge systems, which are based on metaphysical beliefs, tend to view knowledge as much more subjective and, therefore, are not as prescriptive in terms of how people should go about acquiring it. In other words, there are many different ways that we can learn about the world and our place within it.

WHAT DOES IT MEAN TO THINK LIKE AN ARTIST?

Disrupt your status quo and challenge assumptions. Embrace uncertainty. Stand on unfamiliar ground with unfamiliar materials and activities. Creativity, curiosity, and courage can help you fulfill your creative capacity, unleash your curiosity, and summon the courage to act on your ideas. Thinking like an artist will help you learn how to see, think, and play, and to experience a shift in mindset that will enable you to see the world from new and unique perspectives. The books, *The Need for Artistry in Professional Education: The Art of Knowing What to Do When You Don't Know What to Do* and *Lines of Inquiry: Arts-Based Learning*, are valuable additions for building a culture of creativity and innovation. They explore the way artists approach their work and consider what skills and capabilities from artistic practice can be applied to other professions' practices. They also explore methods of inquiry, such as creative facilitation practices using arts-based learning processes to facilitate groups of people for learning and development. Storytelling-as-inquiry is used as a methodology for understanding participants' experiences of being involved in arts-based learning.

The arts-based approach gives a new perspective to looking at situations. It opens your senses to what is going on around you. It enables the creative side to come out and inspire different perspectives. It allows for freedom of collaboration. Arts-based learning is the instrumental use of artistic skills, processes, and experiences as educational tools to foster learning in non-artistic disciplines and domains.

Arts-based learning facilitation is a dynamic and emerging field of practice and has been defined as a pedagogical means to contribute to the learning and development of individuals and organizations through the application of a wide range of arts-based approaches. The application of arts-based learning initiatives and interventions provides people and organizations with a unique way to look at and engage with their organizations or communities. Arts-based learning is not about turning people into artists; however, it does allow professional practitioners in any field to think artfully and creatively about what they do. It offers new insights and different perspectives and provides new approaches for decision-making and solution-finding. It allows mindsets and pre-conceived ideas to be stretched and at times challenged in creative ways.

Creativity lies at the heart of everything artists do. We also tend to resist change rather than embrace it. Every country on Earth is reforming public education. There are two reasons for this: the first is economic and the second is cultural. How will our children take our place and maintain a sense of cultural identity so they can pass on the cultural genes of our communities? In the past, when students went to school, they were kept there with a story: if you worked hard and did well, you would get a job. However, in the process, most of the things that were important about them were marginalized. The current system of education was designed, structured, and conceived for a different age in which education was assumed to be for the rich—not for the public—and definitely not for free. Deep in the gene pool of public education, there are two types of people: academic and non-academic people, smart and non-smart people. The consequence is that many brilliant people think they are not brilliant because they have been judged against this particular view of the mind. If you think of the arts, they specifically address the idea of aesthetic experience. An aesthetic experience is one in which your senses are operating at their peak; in which you are present in the current moment, where you are reasoning with the excitement of the thing that you are experiencing, in which you are fully alive. Our education system should be waking us up to what we have inside of ourselves, not supporting conformity, standardized testing, and standardized curricula.

A culture of creativity is important because it fosters a learning environment that supports creativity. Art making is a practice of presence. When we enter into an experience of art making, we enter an alternative space, both physically and psychologically a different state of consciousness than that of ordinary life. In artistic responding, the dreamer and witness offer artistic responses to the dream story using available materials. These artworks are not literal; by embracing surprises, the arts can be effective in many contexts, including counseling, coaching, psychotherapy, supervision, organizational development, team-building education, and social change. In educational institutions, students are often categorized by labels according to their scores on intelligence and achievement tests. If we emphasize a person's uniqueness, it does not make such categories wrong, but the categories are placed into the background of our thinking, and we are better able to recognize the unique gifts and resources, as well the individual challenges and problems that each person experiences. Categories of all kinds can shape our thinking in strong ways. It is important to remember that human beings make up these categories and that they change over time and across cultures. A large percentage of people obtain and retain knowledge and information visually; people actually SEE what they have heard. As a result, they are encouraged to contribute to conversations and discussions because they feel they have all of the information and knowledge visible, which makes it easier for them to make connections and see the whole picture. Artists and business leaders have many parallels. Both roles involve having a guiding vision and a point of view, formulating an ideal, navigating chaos and the unknown, and, finally, producing a new creation.

Studying the arts can help people communicate more eloquently, and studying in the art world might even hold out the biggest prize of all—helping people become more innovative. Creativity is not a mystical attribute reserved for the lucky few. Creativity is a process that can be developed and managed. Generating innovative ideas is both a function of the mind and a function of behaviors; anyone can put it into practice. Creativity begins with building a foundation of knowledge, learning a discipline, and mastering whole-brain thinking. We learn to be creative by experimenting, exploring, questioning assumptions, using imagination, and synthesizing information. Arts-based activities include drawing, painting, storytelling, theatre improvisation, photography, and poetry—to name a few. Arts-based activities can be used strategically to create safety, build trust, find shared values, and shift perceptions.

Indigenous knowledge is not yet fully utilized in the educational development process. Conventional approaches imply that development processes always require technology transfers from locations that are perceived as more advanced. This has often led to developed nations overlooking the potential in local experiences and practices. Indigenous knowledge is an emerging area of study that focuses on the ways of knowing, seeing, and thinking that are passed down orally from generation to generation. These ways of understanding reflect thousands of years of experimentation and innovation in topics like agriculture, child-rearing practices, education systems, medicine, and natural resource management—among many other categories (World Bank 2007). These ways of knowing are particularly important in the era of globalization—a time in which Indigenous knowledge as intellectual property is taking on new significance in the search for answers to many of the world's most vexing problems: famine, ethnic conflict, poverty, and so on.

The result of all this is that Indigenous knowledge has, at best, been sidelined or looked down upon; but, at worst, been deliberately suppressed or eradicated (World Bank 2007). Universities—under pressure from increasingly sophisticated campaigns by Indigenous people—are now beginning to resolve this, but there are still relatively few schools in the world that make Indigenous knowledge a priority. Educational institutions tend to resist change rather than embrace it. Our attitudes about the arts inform our practice. The arts can help to open and expand a question, to reveal images from dreams, to enhance community cohesiveness, and, especially, to activate imagination. Many practitioners focus upon the arts as a way to experience "flow" (Csikszentmihalyi 1996) or to express and contain emotion and experience. Art making offers opportunities to practice focusing and decision making.

Art making can be a process of inquiry leading to emergent knowledge—knowledge that is new, unanticipated, and unpredicted. Shaun Mcniff (1998) has pointed out that art is not only a way of knowing, but also a research method and an attitude toward inquiry that restores life and imagination to research endeavors. The art-making process addresses attitudes, playfulness, non-judgment, and trusting the process; the artist's way to working with the arts is open and playful. Begin with sensitizing yourself to the materials to be used. Working with the process and products of art making in expressive arts requires the same attitudes of non-judgment that are requisite to any helping relationship. In the arts, withholding judgments can be particularly challenging because of cultural and personal tendencies to judge both ourselves and our art.

STORYTELLING AND THE IMAGINATION

Elder members of Indigenous societies oversee the learning process. In addition to acting as guides to the land and its flora and fauna, they also convey knowledge to younger individuals by telling stories (Dunn 2014). These stories provide wide-ranging information to their listeners, such as how the earth was created, the way in which animals and plants came about, why certain moral rules exist within a given society, and so on. They are imaginative stories that not only portray and celebrate the inherent beauty of the environment, but also allow their listeners to relate to the objects of the story and empathize with them (Dunn 2014). The listeners to the stories imagine what it is like to be the animals in the stories and experience the world from their perspectives. Stories and knowledge may be passed on in many different ways. For example, among Australian Aborigines, information about the landscape is traditionally turned into songs. These songs are then passed down from generation to generation, helping those who know them to get from one place to another. The lyrics of these songs also provide an account of how the world came into being. So when the songs are sung (and sometimes danced), they not only provide a practical guide for getting from one place to another, they also recreate the birth of the earth, simultaneously blending the physical and metaphysical worlds (Dunn 2014). Children learn the song lines from a very early age, carried around on their mother's backs. The result is a unique fusion of imagination, artistic ability, and navigational skill. They owe their skills, which outsiders find so remarkable, to the acquisition of power in this way and to paying close attention to hundreds of stories and accounts of how to behave in different situations (Dunn 2014).

There are two important implications of using these methods of passing on knowledge. First, the need to use one's imagination means that listeners develop a profound bond with the environment in which they live and the rules required to live successfully within it (Dunn 2014). Second, storytelling, song learning, and ritual dancing help to strengthen community bonds, with the younger generation learning from the older and building respect for their knowledge and position in society. This is why many Indigenous societies consider written knowledge to be inferior to spoken knowledge and why some have even rejected the former. They see the advantages of learning empirically and personally as more than making up for the insecure position it places them in when it comes to protecting their knowledge; having said that, it does make them more vulnerable (Dunn 2014).

Unfortunately, many governments and corporations have taken advantage of this vulnerability, realizing that, if you take away the language of an oral society, you also take away its identity and knowledge (Dunn 2014). You are then free to take away its possessions, which are often a crucial part of the society's history and operations. The ideals of arts-based learning are about broadening thought and challenging the brain to process information differently. In terms of how this may affect our professional lives, so much of what we do is "ticking the boxes." By this, I mean that we are simply following a recognized process for achieving an end goal. Arts-based learning challenges this process by encouraging alternative thinking; hopefully, this can have a positive impact on increasing innovation in time.

In this sense, art and culture are indispensable elements of a comprehensive education, the objective of which is to achieve the maximum benefit for and best possible development of every individual, thus enabling them to actively participate in society as constructive members of the community. The ability to capably handle the challenges encountered by each individual and by society, the capacity to deal responsibly with resources and the environment, well-developed interpersonal and communication skills, a well-developed faculty of reflection, and the abilities to learn, to make decisions, and to act competently—all these are educational goals that, when achieved, contribute substantially to a satisfying life (Putz-Plecko 2008). In this context, experience with art facilitates learning processes that are of pivotal importance—for example, the ability to recognize and relate to what is different, or the capacity to develop transitions and interrelations within a heterogeneous group (Council of Europe 2016).

Like science, art can contribute to an overall process of development in society as a result of its view of the world and its approach to creativity (Putz-Plecko 2008). Moreover, cultural education creates a constructive basis for encounter and discussion, for coexistence and cooperation. It is a foundation of general education, not a luxury that may be added when all other educational goals have been achieved.

The function of schools in society is not only to give our children knowledge and skills, but to open up spheres of experience and development in which young people can get to know themselves and become familiar with the world, comprehensively fostering the development of their personalities (Putz-Plecko 2009). Education "in the arts" and education "through the arts" open up access to a more widely defined cultural education and are an essential part of it at the same time. The aim of education must be to promote the full development of the personality, talents, and mental and physical capabilities of each individual child. Cultural education—that is, education in the arts and education through the arts (which means the use of art-based forms of teaching as pedagogical tools in all kinds of school subjects) as examined by Anne Bamford in her systematized and comparative global review written for UNESCO ("The WOW Factor", 2008)—makes an important contribution to the achievement of this aim. It is, in effect, a motor of individual development.

Artistic processes relate to the world "differently." Learning is a creative process. What we learn depends, for the most part, on how we learn: on the learning place and learning atmosphere, the time, the rhythm, and the clarity of the presentation (Putz-Plecko 2009). Learning by means of art-based methods opens up specific spheres of experience and development, an easily recognizable fact that has been examined and described in numerous studies. This form of learning is distinguished by the particular vividness and clarity with which knowledge and ideas can be communicated (Putz-Plecko 2008). Moreover, it promotes a positive understanding of diversity, of different approaches, and of multi-perspective ways of viewing things; for example, by directly conveying the insight that there is more than one reasonable answer and more than one solution to a problem (Putz-Plecko 2008). The demographic changes that schools now have to deal with constructively have made it clear that a homogeneous culture for all simply does not exist, and that, more than ever before, we are being called upon to take this into account (Putz-Plecko 2008).

A sensuous approach lets children discover new worlds and come to grips with them in playful ways. The essential elements are perception and creativity, as well as the enjoyment and adventure of seeing, hearing, trying things out, simulating, playfully transforming, and achieving. It is a matter of providing indispensable and different access to the outer and inner worlds (apart from the cognitive approach through technologies and media; Vienna: Hanser Verlag 1998).

It is a proven fact that we need sensuous experiences in order to develop. Neuroscientific research has demonstrated how thinking is stimulated by the senses, and that creativity requires neuroplasticity. Literacy—as an educational goal and key area of competence—must be more than language as a verbal means of expression and communication (Putz-Plecko 2008). Creative design takes place in many "languages," and schools must provide the time and space for the development of both linguistic and non-linguistic forms of expression. Experimental situations must be permitted or created that will allow young people to make discoveries and develop new things or new modes of performance in their own languages, in their own individual ways, and in their own personal forms of expression. Artistic processes are always search processes that involve seeking individual paths, and, at the same time, they are processes of creative thinking. They bring new understanding of oneself in combination with greater understanding of and new connections with the world, and they produce insight and knowledge in a special way (Putz-Plecko 2008).

Aesthetic education—education in the arts—is distinguished by a specific interaction of cognition and emotion. We know that people who do not learn to deal with emotional intelligence run the risk of developing large deficits in perception, decision-making ability, and the capacity to cope with everyday life and social situations; nevertheless, very little is being done to put these insights into practice (Putz-Plecko 2009). A great number of good intentions are being expressed, but very few political measures are being taken that could facilitate and promote a paradigm shift. Increased effort has to be made to establish synergies between knowledge, skills, and creativity. With few exceptions, educational politics gets no further than paying lip service to these ideas (Putz-Plecko 2008).

Creativity and the ability to innovate are decisive for sustainable economic and social development (Putz-Plecko 2009). At present, we are at the end of the industrial age. The abilities and skills that were needed to safeguard the social order of industrial society are losing relevance (Putz-Plecko 2009). The modern working world is no longer primarily defined by demand, but rather by perpetual renewal and innovation. One of the most essential competences that will be needed in the future is the ability to decide—for the most part independently and without reference to predefined work processes—which possible solution to a new problem is the right one. Knowledge and creativity are the new economic factors. It is not raw materials and machines, capital, or land that analysts say will be the driving forces of our economy in the future. Creativity needs freedom. It is something more than "superficial creativity." But creativity—the ability to create something new—needs to be allowed to grow and develop in its own way and in its own time; it needs patience and faith. Creativity cannot be measured or produced according to scale. It cannot be categorized. It lives on freedom, not commands. (The problem is the systems we have now are enormous categorization and compartmentalization machines that oppose behavior that diverges from the norm.) Creativity requires self-organized people and tolerant, open communities. This is what cultural education aims for. Teachers play a decisive role in awakening and encouraging creative potential. They provide examples in the way they teach and through their personalities. This is why teacher training is so important; teachers must be capable of capturing their pupils' interest and fostering their abilities in the sense described above and, thus, of presenting lessons clearly and competently enough to achieve these goals. At the same time, they must be interested in interdisciplinary approaches and have the corresponding capacity to work cooperatively. As a matter of fact, a large percentage of teachers are in favor of artistic and cultural activities, and a number of them engage in such activities with commitment, resulting in lasting positive effects for everyone involved. At the same time, there is an obvious disproportionality between the personal commitment required on the part of the teachers and the small amount of institutional support they receive in a system where openness to such activities is lacking. Often enough, teachers have to battle all kinds of structural resistance before they can carry out their activities.

Art and educational institutions have an educational mandate. A new learning culture has to be promoted by enabling new learning communities and supporting networks (Putz-Plecko 2009). Cultural institutions, too, need to rethink their roles in connection with cultural education; generally, the production, presentation and preservation of the cultural heritage are placed in the foreground while education takes a back seat (Putz-Plecko 2008). Aesthetic education thus leads to the heart of cultural education. The important thing is to provide a sensuous approach and a playful exposure to art and culture. Other essential elements are thought, communication, and integrative processes. While artistic education starts with the subject, cultural education is dialogical in nature and focuses on the way people deal with their fellow human beings and with the environment. It contributes to people's socialization and strengthens their ability to participate actively in the life of society—at various levels and in a variety of ways.

When we speak of culture and education today, we have to take into account global migrations, worldwide communication networks, international business groups, and the problem of poverty, which concerns all societies (Putz-Plecko 2008). Culture is understood as a process, as the production and exchange of shared meaning. It is in this sense that these art/education projects follow up on the various levels of construction of the so-called "other" and so-called "self."

CONCLUSION

Most people have long accepted some of the fundamental assumptions underlying education. We have assumed that knowledge is a necessary form of intellectual liberation that opens to the individual options and possibilities that ultimately have value for society as a whole.

At one level, knowledge and education appear beneficial to all people and intrinsic to the progress and development of modern technological society. This notion has been used as a means to perpetuate damaging myths about different cultures, languages, beliefs, and ways of life. It has also established Western knowledge and science as dominant modes of thought which distrust diversity and jeopardize us all as we move into the next century. Cognitive imperialism is a form of cognitive manipulation used to disclaim other knowledge bases and values (Battiste 2011). Validated through a particular cultural based of knowledge and empowered through public education, it has been the means by which whole groups of people have been denied existence and had their wealth confiscated (Battiste 2011). Cognitive imperialism denies people their language and cultural integrity by maintaining the legitimacy of only one language, one culture, and one frame of reference (Battiste 2005).

Cultural education thus implies opening up our society by means of art and culture. Open forms of learning and shared creative processes create a space for encounters and for dealing constructively with differences. This space has physical, intellectual, sensual, emotional, and social dimensions. Cultural education cannot simply be prescribed; it requires a new culture of teaching and learning, which

- is open and cooperative both internally and externally,
- focuses on the needs of the pupils,
- is open to innovative, interdisciplinary work, and
- is project-oriented.

It is a common concern and a dynamic process involving parents, pupils, teachers, school administrators, artists, cultural educators, and the societal environment, as well as the industrial, political, and administrative sectors. Stretching boundaries always begins with taking an interest in others. It requires openness to new things and the courage to become involved in something that could develop in unforeseen ways.

Whether it is language, media culture, music, visual culture and art, dance, drama, material culture, rites, forms of thinking and acting, or other aspects of everyday culture that form the basis of cultural dialogue, we are drawing on a treasure trove of languages, ways of articulating, knowledge and experience, history and the present day. Cultural education grows out of learning processes that take the inner differentiations and complexities of culture into account. It lets us experience the learning process with the senses and allows us to internally comprehend how people, under different conditions, have understood the world, interpreted it, acted in it, and changed it in different ways and continue to do so.

It is when resources become scarce and insecurity grows and when a lack of opportunities allows no visions of the future that borders are closed up tight, and there is a tendency to retreat behind hardened constructions of what is one's own and what is alien. Idealization and demonization are frequent side effects of this process. Deviations from what is familiar engender rejection and aggression or are penalized with marginalization. Discussions about identity and culture then often serve the purpose of justifying separation and exclusion. Cultural education programmers cannot work miracles, but they can introduce new perspectives, make restrictive views and actions perceptible, and challenge people to make changes in their habit of thinking, however small (Putz-Plecko 2008). In this sense, they can open up negotiation spaces, promote active debate and, thus, help to develop an accepting culture. Differences should be accepted and constructively negotiated. Every society is permeated with cultural differences; they arise between generations as well as between different communities and sexes, between social groupings, and between varieties of self-chosen affiliations. They develop as a result of people's different backgrounds and reflect different life situations. We carry them inside us. They construct identity, and how we deal with these outer and inner differences plays a role in determining our future.

References

Ahmed, S. (2000). Who Knows? Knowing Strangers and Strangerness. *Australian Feminist Studies, 15*(31), 49–68.

Battiste, M. (2005). *Indigenous Knowledge: Foundations for First Nations.* Sasktoon: University of Saskatchewan. Retrieved April 22, 2015, from Vancouver Island University Website: https://www2.viu.ca/integratedplanning/documents/IndegenousKnowledgePaperbyMarieBattistecopy.pdf.

Battiste, M. (2011). *Reclaiming Indigenous Voice and Vision.* Toronto: UBC Press.

Council of Europe. (2016). Lifelong Learning Platform. *Report on Cultural Education.* Retrieved April 22, 2015, from Lifelong Learning Platform Website: http://lllplatform.eu/policy-areas/cultural-education/report-on-cultural-education-council-of-europe/.

Csikszentmihalyi, M. (1996). *Creativity: The Psychology of Discovery and Invention.* New York: HarperCollins.

Dunn, M. (2014, September 26). *Linking Indigenous and Western Knowledge Systems.* Retrieved April 25, 2015, from The Theory of Knowledge Website: *Theory of Knowledge.net.* http://www.theoryof-knowledge.net/areas-of-knowledge/indigenous-knowledge-systems/linking-indigenous-and-western-knowledge-systems/.

Flavier, J. M. (1995). The Regional Program for the Promotion of Indigenous Knowledge in Asia. In D. M. Warren, L. J. Slikkerveer, & D. Brokensha (Eds.), *The Cultural Dimension of Development: Indigenous Knowledge Systems* (pp. 479–487). London: Intermediate Technology Publications.

Gurteen, D. (1999, February). Creating a Knowledge Sharing Culture. *Knowledge Management Magazine, 2*(5). Retrieved April 28, 2015, from Gurteen Website: http://www.gurteen.com/gurteen/gurteen.nsf/id/ksculture.

Hansel Verlag, C. (1998). Retrieved April 28, 2015, from Sebald Wordpress Website: https://sebald.wordpress.com/category/hanser-verlag/.

Knill, P., Barba, H. N., & Fuchs, M. N. (2004). *Minstrels of Soul Intermodal Expressive Therapy.* Toronto: EGS Press.

Mabule, M. (2016). *Born to Be Different. Blogspot.* Retrieved May 2, 2015, from Mabule Blogspot Website: http://mabule.blogspot.ca/2015/11/developing-our-indigenous-knowledge.html.

Mcniff, S. (1998). *Art Based Research.* London: Jessica Kingsley Publishers.

Meyer, J. P., Allen, N. J., & Smith, C. A. (1993). Commitment to Organizations and Occupations: Extension and Test of a Three-Component Conceptualization. *Journal of Applied Psychology, 78*(4), 538–551.

Murad, F. (2010). *Narratives of Hope in Anti-Oppression Education: What Are Anti-Racists For?* Toronto: University of Toronto.

Putz-Plecko, B. (2008). *Background Report on Cultural Education: Promotion of Cultural Knowledge, Creativity and Inter-Cultural Understanding Through Education.* Vienna: University for Applied Arts. Retrieved May 2, 2015, from Study Lib Website: http://studylib.net/doc/8293849/cultural-education–the-promotion-of-cultural-knowledge--....

Putz-Plecko, B. (2009). *Art and Culture: Key Elements of Education. An European Asset.* Vienna: University for Applied Arts. Retrieved April 29, 2015, from Study Lib Website: http://studylib.net/doc/8293849/cultural-education–the-promotion-of-cultural-knowledge--....

Razack, S. (1998). *Looking White People in the Eye: Gender, Race, and Culture in Courtrooms and Classrooms.* Toronto: University of Toronto Press.

Said, E. W. (1994). *Culture and Imperialism.* New York: Vintage.

Sardar, Z. (1999). Development and the Location of Eurocentrism. In R. Munch & D. O'Hearn (Eds.), *Critical Development Theory: Contributions to a New Paradigm.* London and New York: Zed Books.

Smith, L. T. (1999). *Decolonizing Methodologies: Research and Indigenous Peoples.* London: Zed Books.

Warren, D. M. (1991). *Using Indigenous Knowledge in Agricultural Development.* World Bank Discussion Paper No. 127.

World Bank. (2007). *Indigenous Knowledge Program.* Retrieved April 26, 2015, from World Bank Website: http://web.worldbank.org/archive/website00297C/WEB/0__CO-23.HTM.

PART IV

Technology

Awakening the Seed of Kenyan Women's Narratives on Food Production: A Glance at African Indigenous Technology

Njoki Nathani Wane

Abstract This paper's focus is the role of women in Indigenous Knowledge production, particularly, the use of Indigenous technologies. The study took place among Embu rural women in Kenya. During my research, I noticed the reliance that the women had on Indigenous ways of knowing, yet this knowledge was not part of what I would teach in the academy. By collecting and piecing together the shards of Kenyan Indigenous technologies is to recapture and legitimize a form of technology that is embedded in the cultural psychology and psycho-ecology of a community. This chapter strives to narrate my conversations with the women elders who felt that, I had been away for too long to be trusted that I would pass on this knowledge to the next generation. However, after rituals of mutual crossing, I was able to earn the elders' trust, because I told them that my time in the West could not change the person I was

N. N. Wane (✉)
Ontario Institute for Studies in Education, University of Toronto, Toronto, ON, Canada
e-mail: njoki.wane@utoronto.ca

© The Author(s) 2018
N. N. Wane and K. L. Todd (eds.), *Decolonial Pedagogy*,
https://doi.org/10.1007/978-3-030-01539-8_7

and that the invisible thread that linked us should not be broken because I had a Western education.

Keywords Indigenous · Knowledges · Indigenous technologies · Women · Knowledge production

INTRODUCTION

In 1994, I journeyed to Kenya to explore the role of rural women in knowledge production and my entry point was food processing and preservation. Growing up in rural Kenya, we always had food. I do not remember any time when our family went without food and this was because of the ingenious ways that my parents employed to preserve any food that was harvested. We did not have deep freezers or fridges. What we had was granaries or special shelves that were lined with hanging maize cobs or other forms of grains. During my research on the role of women in Indigenous knowledge production, I could not help but notice that the women still employed unique methods to store their produce.

Collecting and piecing together the shards of Kenyan Indigenous technologies is to recapture and legitimize a form of technology that is embedded in the cultural psychology and psycho-ecology of a community. It is not enough to identify its technical contents. One must comprehend the beliefs, values, symbols, myths, and the social-cultural-cosmological components with which it is associated. In some instances, aspects of these cultures are appropriated without acknowledgement either through written texts or media, and in the process their authentic derivations are subsumed or obliterated. This paper strives to narrate my conversations with Embu rural women elders who have preserved their Indigenous methods of food preparation and food preservation, but who feel that, I have been away for too long to be trusted and that I will pass on this knowledge to the next generation. It is important to note that, I was able to earn the elders' trust through rituals of mutual crossing (see Wane 2014), because I told them that my time in the West could not change the person I was and that the invisible thread that linked us should not be broken because I had a Western education. I would not and could not allow such a travesty. In this paper, my focus is on Indigenous Technologies that many rural Kenyan women relied on for their food processing and preservation. The paper is a reflective piece on what I witnessed during my long year stay with women in rural Kenya. I situate my argument on African Indigenous knowledges.

KENYA—THE LAND OF DIVERSITY

Kenya is a country of startling beauty. Travelling through Kenya one is amazed by the contrasts in landscape, lifestyle and people's beliefs. And it is in Kenya, amid escarpments and scattered lakes, that one finds evidence of man's earliest ancestors. I was born in this land, and came of age within traditions enriched by more than a thousand years of cultural exchange of Nilotic and Cushitic language groups whose lives were shaped by the fertility of the land, or lack of it (Wane 2014). Agriculture was foremost in the survival of peoples whose ethnicity and the type of food production were related, marked by subsistence strategies that exist to this day. This knowledge encompassed not only agricultural practices and methods, but also the social and spiritual. Formed by observation of the natural world, and dictated by the vagaries of the landscape, these peoples nurtured a healthy respect for the land and its creatures, growing food crops for domestic sustenance and often trade. Achieving these ends were integral to traditional practices, with food processing techniques passed down through the centuries to reside in the understandings of women who today engage in agricultural practices much as did their forebears. Their domestic food processing techniques are part of a wider Indigenous knowledge base that regrettably is under threat, first precipitated by colonization, and now by the onrush of globalism and modernity.

AFRICAN INDIGENOUS KNOWLEDGES

When the subject of African Indigenous knowledges is raised we are often asked to define its meaning. To the African, these knowledges are not linear, and their essences are shared by communities that encompass a world view that engages and embraces the totality of being and living. It is this commonality that allows us to examine our understanding of African cultures, politics and economics (Wane 2002). Because African Indigenous knowledges are dynamic and founded on diverse life experiences, a neatly packaged definition is elusive. Indigenous knowledges is viewed as an ambiguous topic that immediately places analysts on dangerous terrain, but I contend that Indigenous discourse is intellectually evocative and warrants reflection, not despite its complexity, but because of it. I also submit that an understanding of Indigenous epistemology provides scholars with another view of knowledge production in diverse cultural sites. These knowledges provide an overt understanding of cultural processes by which information is legitimated and delimited. Learning from Indigenous knowledges through what local communities know and practice can improve understanding of local conditions and provide a constructive context for activities designed to help the communities.

Indigenous knowledges evolve according to need and are developed and modified for efficiency in time-saving. They are cumulative and represent generations of experiences, careful observations and trial and error experiments. The community builds on these knowledges, expands them, and refines them out of necessity and out of a spiritual resonance with their environments (Wane 2003; Semali and Kincheloe 1999). Children watch their mothers at work and try to imitate them during play. When allowed, they carry out tasks under the watchful guidance of their mothers or aunts. Mastery is gradual, the learning processes a daily activity, and training and education is not relegated only to specific times and places.

Rural Women Elders

The rural women elders, the custodians of ancestral knowledge and history are in some wider sense educators whose knowledge has depth and unthinkable scope. Their education is eminently *practical* and *valid*, no matter its marginalization from a Western classroom and Western mores. I had returned to Kenya with the express purpose of gleaning a greater knowledge and understanding of Embu rural women's food processing practices.

During my conversations with my research respondents, I recognized that their deep knowledge of the ecology, nutrition and traditional technologies exceeded that of most government planners, officials and what I have gained through several years of Western education. Even though many of these Indigenous women were proud of my achievements, it was

I who looked at *them* with envy. I realized that I could resolve my internal dilemmas if I could debunk Western-formulated myths and wrestle from almost certain obscurity the Indigenous food knowledge of Kenyan women. This was a significant aspiration, because by so doing I would pass on this knowledge, help prevent the genocide of another cultural knowing, and protect the birthright of future generations.

In this paper, I strive to document women's narratives on food processing technologies without reproducing the epistemic violence that has so often accompanied research practices. It is critical that women's technological knowledge be situated within a worldview shaped by collective understanding and interpretations of their social, physical and spiritual world. The legacy of colonialism, in its discrediting of Indigenous knowledges, has created a gap between two worlds of knowledge production: The Western, and the African. The camera, wielded as an instrument of political and cultural manipulation, is not used to capture this 'other' face of Africa. This Western propensity to ignore African success in favour of focusing almost obsessively on African failure helps us consider how social constructions and reconstructions inform our understanding of representation and portrayal. And the lens through which Westerners view Africa in general, and African women and children specifically, offers distorted images blurred by colonial legacies, patriarchy, economic imbalances and specific political agendas.

I am responding to Ivan Van Sertima's (1984) message, in *Black Women in Antiquity*, who states that there is a need for African peoples to unearth the knowledges buried beneath the weight of years of oppression under the colonial and neocolonial rule. We must acknowledge, revisit and respect the many sites of ancestral knowledges in order to correct the distorted imaging, identities and histories of African women.

Witnessing the Knowledge Creation

As I have indicated elsewhere (Wane 2014) Indigenous knowledges has not yet gained complete acceptance in the academy. This is clearly demonstrated by Sittirak (1997) close observation of her mother's life. Sittirak mother knew the importance of environmental conservation. Instead of using aluminum foil or plastic wrap, her mother would use banana leaves. Through observation of her mother's life, Sittirak was able to "make critical connections between the theoretical and practical issues related to the environment and development" (Sittirak 1997, p. 125). She then continues to question "why her mother's voice remains unheard" (p. 125). Sittirak's conclusion is that, it is due to the "Western knowledge producers' reluctance and adamant refusal to acknowledge that indigenous women are reservoirs of information, with abilities and knowledge gained over generations of experience" (p. 125). She continues to state "The practical, not the theoretical is their (rural women) dominant criterion. Either something works or it doesn't" (p. 125).

Among the rural women knowledge of the land is unquestionable. When it comes to taking care of the land, the women know which plants to keep together and which ones to be separated. The women know what parts of the plants sustain what. For instance, the Embu rural women knew that planting onions, coriander and lemon grass would drive worms away because of the scent these plants produce. The scent from these plants serves as a natural pesticide. The women also knew that planting Jerusalem and Sodom's apple around the farm will keep snakes away. In addition, they know that that maize loosens and oxygenates the soil, and that certain beans roots, while others loosen it. This knowledge is priceless, but it is being lost, unless as Ivan Van Sertima (1984) says, we document it.

A place for teaching Indigenous technologies, understood as a counter hegemonic discourse, is essential to the survival of African communities. If women cannot reclaim their knowledge they will gradually lose both knowledge and identities. Full recognition and acknowledgement of these women's knowledge is a reactionary process and a proclamation that they have their own knowledge, their own identity and do not need to be defined by the Western world or other cultures. We have to accept the universality of knowledge and the globalization of the technologies themselves by the outside world.

During my research, I was riveted by the women's activities. They were seated on short, homemade stools, surrounded by everything necessary for their work. I never witnessed a single moment when their hands were idle. With rapid movements they would reach for the firewood drying from the heat of the fire in a *rutara* (raised platform) and would take water from the earthen pot at their side and dip utensils in an urn to soak.

It is these elders who taught me the importance of appropriate technology, reminding me that one's culture is central to self.

FOOD PRESERVATION, TRADITION AND CHANGE

Boxall shows that in countries where farming is predominantly at the subsistence level, an estimated 60–70% of cereal and grain produced is retained at the farm level. Farmers must conserve enough grain to feed both family and livestock from one harvest to the next. Storage methods have evolved according to local customs, social and economic conditions and have become associated with certain varieties of grains. Newcomers to a community unfamiliar with post-harvest activities may erroneously conclude that these methods are inefficient and wasteful when compared to mechanized processes. However, a closer examination would reveal that costs are often lower when compared to labour costs.

A variety of traditional storage systems have evolved in developing countries to satisfy local requirements. Although basic concepts may seem similar, construction methods and appearance vary according to local conditions and customs. In most cases storage facilities are constructed using plant materials, mud or stone. Facilities are often on above ground platforms. If the harvested crop does not require a large storage facility, containers such as baskets, clay pots, gourds or drums are used.

According to rural Embu women, methods of preservation range from drying to the use of pesticide, ashes, marigold, hot pepper, onions, cow dung, garlic, trash and aromatic leaves. The selected preservation procedure depends on the preservatives being used, and especially the type of food it is to be used on. The success of the procedure depends on who manages the crop from harvesting to storage. As Ciarunji (one of the participants interviewed) stated, "What I use depends on the crop. For instance, I use marigold for maize, and for beans and potatoes I use ashes." According to Mama, another participant, "I use whatever is available. Sometimes ashes, other times paraffin or soil." According to these women, there is no specific preservative that is used in preservation. The ranges are wide.

According to Ciarunji, what matters is planning. As she explained: "if you want to use ashes, start early to collect it, and whenever you are not busy, clean it, sieve it so that when your food is ready for preserving, you scoop ashes and mix it with your food...just like you do with pesticide".

Warue, another participant agrees with Ciarunji.

You plan ahead of time and harvest your ashes and get them ready for your food...you cannot leave it up to the last minute [or] you will not have time to prepare and use it. Marigold used to grow everywhere, but this is not the case anymore because most of the land has been cultivated... this is what it means to progress... to destroy everything we depend on... our herbs, medicines and preservative.... Is this what they teach you out there... to destroy anything that is on your path? You have been gone for long and you are asking us what happened... we should be asking you? What happened...?

Cucu, another participant, was as upset as Ciarunji.

I see no sense in using foreign powders to preserve food. Do you know what is in those packets? I cannot stand the pesticide scent, nor eat any foods preserved with it....I do not understand why the younger generation "rushes" into things blindly?

Although I had the answers to their questions, I did not want to let them know that the destructions of their preservatives were part of the colonial and neocolonial machinery. These ideologies are like trucks that have no brakes and they destroy anything they find on their way as they finally end up crashing a tree, people, building or in a gushing river that takes it away. To change the topic, I asked Cucu about cooking and this is what she said:

"Cooking is not difficult... to preserve food these days is a problem... We never had weevils. We started having problems with the weevils soon after the Second World War. After harvesting, we kept everything in Miruru (granaries) or hung the maize on branches. This food could stay there until the next season. Those days, the seasons never changed...we knew when it would rain so we did not fear that the food would be destroyed if it were left on the shamba (farm) to dry. These days you cannot do that. If it is not the weevils, the moths, or the worms, the rains will destroy your food or it will be stolen from your shamba".

From Cucu's narratives, the rural communities are facing many challenges with food preservation. With regard to weevils, it is suspected that these pests were either brought back by soldiers returning from the war, or borne on the various crops introduced by the colonizers. The true origin of their infiltration, however, remains unclear. Despite all these challenges, I still wanted to learn and to find out what they knew in terms of Indigenous knowledge on food preservation. And Wachiuma, a participant said:

...the procedure for preparation of the preservatives depends on the crop and season... For instance, to prepare a solution for spraying vegetables takes about 5 to 10 days. Plants beginning to flower are cut into pieces, soaked in a drum that is half-filled with water and left for 5 to 10 days, along with occasional stirring. After the fifth day the decomposed plants are removed and used as mulch for vegetables. The rest of the solution is strained or filtered then diluted with soapy water and sprayed on the vegetables. This solution is effective in repelling aphids, caterpillars and flies. Similar steps are followed in preparing a solution from hot pepper, onions or garlic.

Ciarunji, on the other hand explained:

"...to preserve dry foods we use, cow dung or ashes, I create layers of these two and in between I place the dray foods. Also, preservation of unshelled maize or millet requires different techniques. ...I make a bed of marigold for the crop, then use more of the repellent as a cover. I use my own judgment...I learnt all this by observing and imitating my mother".

As I listened to these instructions, I was amazed at the attention to detail. Every so often the women asked me whether I was following what they were saying and I would nod. But things have changed for these women. It has become difficult, if not impossible, to preserve their foods for more than a week. One of the elders recalled with nostalgia: "When I was growing up all foods could be preserved for a week or more; but not these days".

Although one cannot be definitive, it is probable that the struggle the rural women have in keeping food for a week without it rotting could be attributed to environmental changes and the use of modern, chemically induced approaches to planting. Most of the new varieties are unable to withstand current temperatures, and more so if there is no refrigerator. Air pollution and the destruction of the Ozone layer may imply that the weather changes have created conditions unlike in the past. Since most homesteads today do not have modern refrigeration facilities, the women must prepare their meals on a daily basis. Any leftovers are served the following day to avoid spoilage and wastage.

Food processing practices help maintain and reinforce a coherent ideology of the family throughout the social structure. During harvests, women's and men's skills are complementary. their organizational and managerial skills quite evident as they take charge of food processing.

Women's knowledge of the land and weather conditions, coupled with their silent communication with nature, gives them a superior position in food processing. Without relying on radio forecasts, they ascertain whether the day will be appropriate for harvesting, threshing or winnowing the grains. This is achieved by standing on the threshold of their homestead and taking in the totality of their environment to determine the day's suitability. Indigenous Kenyan women, therefore, play a central role in food production, processing and storage.

Some Thoughts

The introduction of modern technology has made traditional food processing practices among Embu women into something that is outmoded. In fact, most scholars in the 'modern world' view such practices as relics of the past. Consequently, there is the argument that Third World countries and rural women need a "technological fix" or "appropriate" technology that will increase efficiency and output. Unfortunately, most of the "appropriate" technologies as conceived by the West and marketed by the multinationals, is gender-blind and do not meet the real needs of women targeted as "beneficiaries." The reasons for this revolve around knowledge of production cycles, accepted forms of knowledge, the exploitative capitalist system and marginalization of the "voiceless." The issues of firewood and water provision are examples of the effects of "appropriate" technology.

Although most of the elders did not object to modern technology, they were concerned about the complete negation of Indigenous technologies. According to Cucu:

> Development is good, but then everything that was old should not be destroyed. We should try to keep some of our good old things, foods and tools.

Rwamba complained about environmental pollution and the destructive element seemingly inherent to modern machinery. There is a *spiritual* aspect that modern technologies lack. They are impersonal and have no intrinsic qualities beyond what they are used for. Indigenous technologies are made by hand, they are personal, and the food prepared using them is achieved by touch, smell and judgment. The women bring their *selves* to the preparation, knowing that what they are doing is for the family's health and sustenance. There is pleasure in the preparation, a giving and a sense of achievement. Thus, we also see the intrinsic *spiritual* aspect that infuses food preparation.

With modern technologies Ciarunji felt that she could not identify with the process of adding value to food. She spoke of the disadvantages of using a tractor. When women agree to have their land ploughed by tractor they give up the handling of the soil to something foreign. The machines uproot any plants, interfere with soil composition, and most importantly, the women lose the sacredness of working with nature. Listening to the elders in the village, I knew that their deep knowledge on land was more complex than I had imagined. Many implementers or policymakers do not think twice when introducing supposedly new technology. For instance, I still remember a conversation I had with some of the elders about the introduction of a cooking stove. The stove was standing, and one required to stand while cooking. For someone who has never lived in a rural setting would welcome the innovation. However, the women felt the multi-purpose use of a fire was ignored by an introduction of a standing stove that had gas fire coming out of the top of the stove. The gas flames according to the women cannot be used to warm their feet. Also, the traditional way of food preparation is done while someone is sitting on a three-legged stool. In a nutshell, these innovations were problematic and the women felt, it was simple courtesy to consult them before introducing equipment that did not relate to their lifestyle.

Cucu cannot see the logic of abandoning Indigenous technologies and then spending money on items such as tea leaves, coffee, maize flour and salt. According to Cucu, nature has provided for the needs of mankind, but sadly this is ignored in the name of modernity. Although modern foods have become a part of the rural Embu diet, Cucu still believes that traditional food is far better than what modernity can offer. "What we had was good. We had herbals growing behind our houses, our streams were unpolluted...today, our steams have dried up and the water is full of pesticides."

CONCLUSION

The elder raises many issues connected with modern technologies that are borrowed wholesale from the West without there being any consideration of the local situation. With Indigenous knowledges, the kitchen is not simply a place for cooking, but also a site to produce and share knowledges. The farms are sites of learning and not just to produce food for consumption. Therefore, the settings and the type of technologies utilized in these spaces should not be shunned as outdated and outmoded. In fact, it is the warmth fire provides that makes the kitchen a gathering space, a space which allows the sharing of Indigenous knowledges through its many forms of transmission. It also becomes the environment to impart moral lessons and discipline to the young ones.

Unfortunately, the introduction of the modern ways of living was enacted without local sentiments, culture and traditions being taken into consideration. Not surprisingly, when the modern stove was introduced in central areas in Kenya and Sudan, local rural women rejected it. Such resistance to changes in technologies was not centred in the ignorance or foolhardiness of these local women, but because the product was unsuitable for local use. Many of the women said that when you buy foods from the shops, you lose something in the process and many of them could not identify with the foods.

References

Semali, L., & Kincheloe J. L. (1999). *What Is Indigenous Knowledge? Voices From the Academy* (Chapter One). New York: Falmer Press.

Sittirak, S. (1997). *The Daughters of Development: Women and the Changing Environment*. London: Zed Books.

Van Sertima, I. (1984). *Black Women in Antiquity*. New Brunswick, NY: Transaction Books.

Wane, N. N. (2002). African Women's Technologies: Applauding the Self, Reclaiming Indigenous Space. *Postcolonial Journal of Education, 1*(1), 45–66.

Wane, N. N. (2003). Embu Women: Food Production and Traditional Knowledges. *Resources for Feminist Research RFR Journal, 30*(1–2) (Spring/Summer), 137–149.

Wane, N. N. (2014). *African Women's Voices on Indigenous Knowledges on Food Processing Practices Among Kenya Rural Women. A Kenyan Perspective*. Toronto: University of Toronto Press.

CHAPTER 8

Role of Latent Local Technologies and Innovations to Catapult Development in Kenya

Njiruh Paul Nthakanio and Eucharia Kenya

Abstract The domination of "colonization shadow" may have reduced the manifestation of young indigenous technologies and innovations that with minimal value addition could help local communities overcome many challenges. Rediscovery of these technologies can bring about wealth and well-being to the local people who are also the inventors. Some of these technologies have either been suppressed or picked up by colonizers to the disadvantage of local inventors. This chapter discusses the useful, locally found technological resources that have not helped local communities but sometimes fetch millions of dollars elsewhere. This knowledge is expected to bring about rediscovery and decolonization so as to use the technologies to improve local lives. In this aspect decolonization is necessary in many sectors of the economy such as medicine which failed to take off from herbal- to industrial-based

N. P. Nthakanio (✉) · E. Kenya
University of Embu, Embu, Kenya
e-mail: nthakanio.paul@embuni.ac.ke

E. Kenya
e-mail: kenya.eucharia@embuni.ac.ke

© The Author(s) 2018
N. N. Wane and K. L. Todd (eds.), *Decolonial Pedagogy*,
https://doi.org/10.1007/978-3-030-01539-8_8

pharmaceutics. For instance, Kenya is the home of over 1100 species, many have medicinal value. While such herds are condemned at "home" as illegal herbal concoctions, they are glorified in other countries as medicine and food supplements. Today, many Kenyans import such medicine and food supplements at unaffordable prices as disease continues to bite. The conclusion is that there are a number of unexploited indigenous technologies and wealth that have remained dormant due to colonized minds and with little decolonization they can earn wealth that can increase wellness and improve livelihoods for the growing population in Kenya.

Keywords Decolonization · Indigenous technology · Wealth · Wellness

INTRODUCTION

Colonization came with western culture that greatly affected technological advancement. It stifled local technologies and cultures. Intrusion of colonization brought about a paradigm shift including foods, medicine, housing and all types of western inclination. Colonized nations were forced to drop their cultures in favour of those colonizers. In this regard, the colonized will mainly refer to Kenya. The result of colonization is that various traditional technologies were abounded but became stunted. History tells us that the aim of colonization was to obtain raw materials for the growing industries in Europe but never to enrich the culture or technology of the colonized. Many technologies that had evolved around industrial revolution and thereafter were strongly entrenched in Kenya.

Since the earliest days, human kind survived through some research that led to innovation that enabled them to improve their living style. Some of the earliest known technologies were for acquiring of basic survival skills like acquisition of food and shelter. In the early days, means to procure food was basically obtained through cultivation and hunting. Therefore, technologies to enable cultivation and hunting were developed while shelters were dependent mainly, on local architectural designs.

Technological development can be demarcated into timelines such as; ancient, civilization, the modern and super modern civilization.

Ancient technologies seem to have been geared for basic, crude and simple weaponry. The regional (e.g. Europe or Africa) disparity was not very conspicuous. However, later development brought bigger differences among regions. This depended on how nations exploited their indigenous knowledge, opportunities and wealth. Civilization followed a pattern of technological development. For example, Egypt and irrigation, Greece and the literary advancement, Babylon and farming, Ghana and gold mining among others. A shift in paradigm was realized with the coming of the Industrial and Agrarian revolutions that led to economic transformation in Europe and America. There followed scramble for resources around the world, including Kenya, and this is what led to "colonize the people and reap off the resources". Some regions prospered, however, some were left behind like Africa. Many technological leaps were seen in the eighteenth century that included the discovery of the steam-powered engine in 1712, inventory of Power Loom machine for weaving cloth in 1765 by Hargreaves 1812 while in 1876, Alexander Graham Bell invented the telephone (*New-York Tribune*, 1877). As Europe was undergoing technological change, local technologies in Kenya and rest of Africa were subjugated by colonization.

CONSUMER OF TECHNOLOGY

During colonization and after independence, many colonized countries continued to import technologies under the program of technology transfer. Mainly linear model (Bradley et al. 2013) of technological transfers has been used to get technology from colonizers to colonized. This is a situation whereby technology is transferred the way it was invented from the university scientist to the end user (Siegel et al. 2004). This is a traditional model of the technology transfer process that was adopted by colonialist. During colonization the western nations came with their ready-to-use technologies and later transferred them to local people. A kind of Europe in Kenya was established. The effects of this are that the new technologies were forced on the local people. The transfer may have been done but they were never completely adopted. This could be the reason why some technologies died with leaving of the colonialist. A double loss was realized, local technologies were suppressed and the western ones started failing after colonizers left. Even if they were to work there was a cost tied to it especially where patents were involved, such as in books printing and pharmaceuticals among others.

TECHNOLOGICAL SUPPRESSION

Intrusion of western-based technologies into African continent suppressed non-competitive (often times "superior") African technologies. The result was that many useful African technologies became latent and little or no research was put into them. These technologies included; medical and surgery, mining, pharmacy, engineering, food processing, weaponry, industrial chemistry among others. Speed of local technological latency or decay was increased by the fact that many of the African technologies were never documented but the knowledge was passed from generation to generation by verbal communication. For example, the Kamba people of Kenya used poisoned arrows during war where a chemical, locally called "*uvai*", was put on the arrow head (Kieti and Coughlin 1990). These were and are biological weapons, which could have been developed to a weaponry industry. Unfortunately, the chemical formula and product itself have never been locally documented as an innovation. In apiary, Kenyans used to make bee hives by tunneling a log of tree which was good to create a natural roof to prevent water from entering inside it in rainy times (Affognon et al. 2015). However, this was abandoned in favour of western box-like beehives. The Kenyan blacksmiths never transformed themselves to modern manufacturing. Local building materials, architecture, engineering, agriculture, medicine, literature and many other technologies went into limbo. Development of medical field seems to have been more affected and for this reason, medical industry will be singled out for a bit more exploration.

MEDICAL ADVANCEMENT

All cultures worldwide have long history of medicine practices. By 2003, about 3.5 million people depended on herbal medicine and 80% were in Africa (WHO 2003). Herbal medicine has a very important role in healthcare, and in remote areas it is the major source of medication (Dery et al. 1999; UNESCO 1998) such that some Governments have encouraged their use (Marshall 1998). Besides, the original standardization of scientific medicine, a process known as Pharmacopoeias was first based on herbal products (Farnsworth et al. 1985).

HERBAL MEDICINE OR WITCHCRAFT

Herbalists are found in all parts of the globe including America, Europe, Asia and Africa (Kung'u et al. 2006). However, wherelse the West transformed their herbal into pharmaceutical industries that in Kenya stagnated. When the western culture infiltrated, African medicine was viewed with a lot of suspicion, and it was associated either with witchcraft or something close to it. However, if any item was proven to be valuable to the west it was referred to as herbal medicine. African-based natural products were disregarded at the discretion of the western recipients. In situations where the herb served the colonialist well, they promoted and referred it as a natural product or raw material for drugs making. Things like propolis, *Prunus afrcana* products and artemisinin are some of herbs used to produce medicine at industrial level (WHO 2009). Chances are that penicillin and quinine, all herbal products could have been rejected if the discovery came from an African. It is worth noting that at one time quinine and artemisinin (Codd et al. 2011) were major drugs to treat malaria. Other plants like neem oil have been used as mosquito repellents for a long time (Liu et al. 2009).

CHANGING PARADIGM IN HERBAL MEDICINE

In Kenya, the value of herbal medicine has continued to dwindle due to human activities. On the other hand, in the Western and Eastern worlds, plants are being developed to be natural synthesizers of vital herb like products such monoclonal antibodies (Murphy 1907) that are used in vaccines, alternatively plants are put into use as biofactories of biopharmaceuticals (Horn et al. 2004). Some of the plants used in biopharming are tobacco (*Nicotiana* spp) to glasshouse-grown Arabidopsis, lotus and moss among others. In the tropics, Kenya included, there is a lot of exploited "natural capital". The western countries have turned to Africa to bio-prospect with a hope of getting new products. Indeed, they have often been discovered and sometime they are pirated. An example is where some bacterial materials that generated millions of dollars were pirated from Kenya (*The Guardian* 2004). Colonization may have added value to some of the local products but the benefit may not have been felt by the indigenous population. For example, Neem and Aloe Vera are widely used to make products like toothpaste and soaps but by large colonial multinational companies like Uniliver. The paradox is that now the major suppressors of the indigenous knowledge are not the westerners but the local people with the western education. Today a number of plants around the globe are highly regarded for medicinal value, among them are; Mayapple or American mandrake, or mandrake (*Podophyllum peltatum* Linnaeus) (Maqbool 2011) found in American, Cape bush-willow (*Combretum caffrum*) found in South African and Pacific Yew or Canadian Yew (*Taxus brevifolia*) found in Canada have anticancer properties and some research may be going on to make medicine from them (Srivastava et al. 2005). However, very little is going on in Kenya to value add herbs like Red Stinkwood (*Prunus africana*) that is also reputed to have anticancer properties.

THE NATURAL CAPITAL

Herbal market is undergoing a global transformation, turning it to be quite a profitable natural capital business. Local practitioners will need education so as to benefit from this growing trade. The market for herbal products was estimated to be US$62 billion by 2006 and it is expected to be US$5 trillion by the year 2050 (Singh 2006). There are about 7,227,130 species in the world out of which 374,000 are plants. About 10,000 species are listed as valuable medicinal plants in Africa (Christenhusz and Byng 2016). Kenya houses over 1100 species (Teel 1985). Demand of some of the herbs in both local and international markets has been in the increase, like in Kenya harvesting of East African Sandal wood was put under presidential ban in 2007 to prevent over exploitation (Taylor 2010). Sandalwood is exploited for its essential oils used in perfumery. The heartwood of the trunk, main branches and roots contain an essential oil.

LATENT INDUSTRY IN AFRICA; EFFECT OF COLONIZATION

Many latent technologies in Africa have adversely affected industrialization. For instance, increasing the value of medicinal and aromatic plants has not attracted commensurate research and business attention locally. There is a bit of publication on value of medicinal and aromatic plants including the works of John Kokwaro in his book Medicinal Plants of East Africa (Kokwaro 2009). However, a lot of knowledge has not locally been transformed to commercial products. Some of the major herbs of commercial interest include; anti-malarial such as African Wormwood (*Artemisia afra*), Quinine, and quinidine. Others include

Aloe vera, Red Stinkwood in English (*Prunus africana or in Kiembu in Kenya it is called Mwiria*) (Komakech et al. 2017).

Colonization has had a negative impact on indigenous industries in general but health and pharmaceutical growth suffered most because any of its products were labelled as African medicine or witchcraft. This was worsened by the fact that many of the industries had no formal documentation. Medical education was called traditional knowledge and it was never given a chance to advance. This made scholars and users shun herbal medicine for a long time. Thus, colonialist succeeded in making Africans believe that local healers were infective and inferior and thus reducing competition with their medicine. This stifled research in African medicine and pharmaceutical studies. A lot of knowledge has been lost because subscription was by word of mouth and as knowledgeable people die so too does the knowledge. So far about 7000 single compounds that are used in modern medicine have been extracted from medicinal plants. This development happened in the west but not in Africa. This could have been some of the dark ages in African studies. The major compounds that have made impact are sweet from wormwood (*Artemisia annua*) leaves for anti-malaria parasites, Aspirin from the Willow tree and Quinine from Cinchona bark. *Artemisia annua* shrub has been used for traditional treatment in China and now it is a major raw material for medicinal-based industry (Dharani et al. 2010). Also, the Neem tree in Swahili, known as *Mwarubaini*, has a number of uses including anti-plasmodial properties (Bashir et al. 2015).

At present, about 25% of the prescribed medical drugs in the developed countries are plant based, but it is as high as 75% in the developing countries (Sanjoy and Yogeshwer 2003). However, herbal and aromatic products industry has been growing steadily in Asia and India. For many years, many drugs have been extracted from plants. Africa can follow the example of China and India in the commercialization of various food and herbal products. For example, *Aloe vera* is today grown from tissue culture at Jomo Kenyatta university of Agriculture and Technology (Kenya) because there are many products made from it.

TECHNOLOGY OF FOOD PROCESSING

In food, there are many technologies that have not captured the eye of the modern researcher. A good example is brewing industry where alcoholic drinks from the west are over celebrated while sometimes tastier drinks, like beer made from honey, sugar cane, banana, sorghum, maize, rice among others have been neglected. Modern breweries over concentrated in the use of barley but today the brewers have discovery the power of sorghum in brewing, and it is gradually replacing barley. In fact, it is being called a "discovery" despite being used in Kenya for many years. In spite of this, local brewers who use it to make beer are arrested and their product destroyed and labelled an illicit drink. This is a suffocation of innovations and entrepreneurship, instead the brewers need to be assisted to properly obtain the requisite standards of their products.

Colonization opened doors to big multinationals to compete with smaller African economies. Although it brought a healthy competition, which is good for a business it suffocated small non-conventional African industries. Many times, the multinational demanded protection from competition and this stunted the growth of small local companies. For example, coffee board placed a law to prevent traditional roasting of coffee beans, however, the same law does not persecute companies like Dolman coffee or Java Coffee groups. Traditional foods like the potato, when roasted on fire is "poor man's" food and when grilled in an oven it is considered high value food or when fried it becomes French fries that is worth millions of dollars. African technology never developed from fire to oven grilling and thus the value of end product and technology itself has remained archived. Researchers and innovators need to de-archive the indigenous technologies and enhance their competition with colonially introduced technologies. If not done well, the same colonial master will pick it, modify it (technology) and patent it. A good example is the patenting of traditional Kenyan bag called *Kiondo* by the Japanese. In horticultural production, when a flower is planted in the precincts of a house it becomes an indigenous ornament or wild plant and when a western company picks it and plants it in a farm it then becomes a cash crop.

Education industry was not spared by the effects of colonization. In Kenya there was a lot of knowledge in Biology, Physics, Chemistry, Astrology, Agriculture, Medicine, Languages among others but no detailed research and development was recorded before colonization. The consequence is that people migrate from Kenya to Europe to study African History or Geography that could have better been studied locally. There are many situations, when students leave African to European nations to study tropical diseases, African culture, Tropical Botany, Tropical soils among others while real samples are in the land they are leaving. This is not because they want diversified views, but because study opportunities are not available in Africa. Where is the missing link? What happened? The African performing art is dying and left only for tourists who just pay a token to the performers. This is why you hear of Russian Ball dance, Jazz music but you never hear of "Masai Performing Arts" despite being a major tourist attraction. Colonialist relegated African entertainment industry as traditional and backward. The result is that good performances and music have disappeared as composers perish.

DECOLONIZING RESEARCH AND TECHNOLOGICAL DEVELOPMENT

Technological innovations bring about development and advancement. This can only come through local research and innovation. However, over reliance on technological transfer brings about technology enslavement. Colonization succeeded in technological enslavement. There is a need to rediscover the technologies that can bring about more knowledge and wealth to indigenous Kenyans. This will enhance exploitation of wealth that has gone unused or undiscovered due to colonization of mind. The objective of this script is to identify useful resources that have not been exploited locally to help indigenous people but quite often they have been used to earn millions of dollars outside Kenya. Ultimately, it is expected that this will bring about rediscovery and decolonization so as to research and use technologies to improve local lives. For example, many medicinal herbs can be developed into herbal farming and eventually biopharming and also local brewers can be converted to standardized beer manufacturers.

CONCLUSION

Technological latency and development in Africa was significantly worsened by colonization. Quite often the colonizers picked and developed some technologies for their own use. Ultimately, the indigenous Africans' technology was subjugated hence innovations and any final product and the benefits that accrued from them. For example, in herbal medicine, the pharmaceutics have been picking up local herbal product and synthesizing a chemical version of the plant compounds and overtime, the use of herbal medicines declined in favour of the synthetics ones. However, interests on African indigenous knowledge are rekindling. Africa content contains a lot of unutilized indigenous knowledge and resources such as many unstudied flora and fauna that can be used to develop products to create wealth for the indigenous people. For instance, pharmaceutics are bio-prospecting to find new drugs while local fabrics and textile like *kiondo* and *kikoi* are targeted by industrialists with an intention to patent them. Documentation of indigenous knowledge is needed to enable its transmission, improvement and its use in production of goods and services for people of Africa. This will rekindle African science, technology, innovation and knowledge management using modern technology. Developed countries may have technology but with limited unexplored knowledge. However, Africa has got indigenous knowledge that can be catalogued and coded into public database for research and development.

REFERENCES

Affognon, D., Kingori, W. S., Omondi, A. I., Diiro, M. G., Muriithi, B. W., Makau, S., & Rainam, S. K. (2015). Adoption of Modern Beekeeping and Its Impact on Honey Production in the Former Mwingi District of Kenya: Assessment Using Theory-Based Impact Evaluation Approach. *International Journal of Tropical Insect Science, 35*(2), 96–102. https://doi.org/10.1017/s1742758415000156.

Bashir, L., Oluwatosin, K. S., Adamu, Y. K., Ali, A. J., Maimuna, B. U., Eustace, B. B., & Blessing, U. A. (2015). Potential Antimalarials from African Natural Products: A Review Horn, ME, Woodard, SL. *Journal of Intercultural Ethnopharmacology, 4*(4), 318–443. https://doi.org/10.5455/jice.20150928102856. www.jicep.com.

Bradley, S. R., Hayter, S., & Link, A. N. (2013). *Models and Methods of University Technology Transfer Department of Economics* (Working Paper Series). University of North Carolina, Greensboro, 13-10.

Christenhusz, M. J. M., & Byng, J. W. (2016). The Number of Known Plants Species in the World and Its Annual Increase. *Phytotaxa, 261*(3), 201–217. Magnolia Press.

Codd, A., Teuscher, F., Kyle, D. E., Cheng, Q., & Gatton, M. L. (2011). Artemisinin-Induced Parasite Dormancy: A Plausible Mechanism for Treatment Failure. *Malaria Journal, 10*, 56. Open Access. http://www.malariajournal.com/content/10/1/56.

Dery, B. B., Otsyina, R., & Ng'atigwa, C. (1999). *Indigenous Knowledge of Medicinal Trees and Setting Priorities for Their Domestication in Shinyanga Region, Tanzania*. Nairobi: ICRAF.

Dharani, N., Rukunga, G., Yenesew, A., Mbora, A., Mwaura, L., Dawson, I., & Jamnadass, R. (2010). *Common Antimalarial Trees and Shrubs of East Africa, A, Description of Species and a Guide to Cultivation and Conservation Through Use* (pp. 1–100). The World Agroforestry Centre. ISBN: 978-92-9059-238-9. http://www.worldagroforestry.org/downloads/Publications/PDFS/B16781.pdf.

Farnsworth, R. N., Akerele, O., & Bingel, A. S. (1985). Medicinal Plants in Therapy. *Bulletin of the World Health Organization, 63*, 965–981.

Horn, M. E., Howard, J. A., & Howard, J. A. (2004). Plant Molecular Farming: Systems and Products. *Plant Cell Reports, 22*, 711–720. https://doi.org/10.1007/s00299-004-0767-1.

Kieti, M., & Coughlin, P. (1990). *Musyimi the Hunter Kamba—Fables and Legends a Traditional Kamba Story, The Wisdom of Kamba Oral Literature.* Nairobi: Phoenix Publishers.

Kokwaro, J. O. (2009). *Medicinal Plants of East Africa* (3rd ed.). Nairobi: University of Nairobi Press.

Komakech, R., Kang, Y., Lee, J. H., & Omuja, F. (2017). *A Review of the Potential of Phytochemicals from Prunus africana (Hook f.) Kalkman Stem Bark for Chemoprevention and Chemotherapy of Prostate Cancer.* Hindawi Evidence-Based Complementary and Alternative Medicine Volume 2017, Article ID. 3014019, 10 Pages. https://doi.org/10.1155/2017/3014019.

Kung'u, J. B., Kivyatu, B., & Mbugua, P. K. (2006). Some Medicinal Trees and Shrubs of Eastern Africa for Sustainable Utilization and Commercialization. In B. O. Awour, D. Kamoga, J. Kungu, & G. N. Njoroge (Eds.), *Challenges to Utilization and Conservation of Medicinal Tree and Shrubs in Eastern Africa* (pp. 13–25). http://www.worldagroforestry.org/downloads/Publications/PDFS/B14816.pdf.

Liu, N. Q., Van der Kooy, F., & Verpoorte, R. (2009). *Artemisia afra*: A Potential Flagship for African Medicinal Plants? *South African Journal of Botany, 75*, 185–195.

Maqbool, M. (2011). Mayapple: A Review of the Literature from a Horticultural Perspective. *Journal of Medicinal Plants Research, 5*(7), 1037–1045. http://www.academicjournals.org.

Marshall, N. T. (1998). *Searching for a Cure; Conservation of Medicinal Wildlife Resource in East and Southern Africa* (p. 12). Cambridge, UK: Traffic International.

Murphy, D. J. (1907). Improving Containment Strategies in Biopharming. *Plant Biotechnology Journal* (2007), *5*, 555–569. https://doi.org/10.1111/j.1467-7652.2007.00278.x, http://onlinelibrary.wiley.com/doi/10.1111/j.1467-7652.2007.00278.x/epdf.

Sanjoy, K. P., & Yogeshwer, S. (2003). Herbal Medicine: Current Status and the Future Asian Pacific. *Journal of Cancer Prevention, 4*, 281–288.

Siegel, D. S., Waldman, D. A., Atwater, L. E., & Link, A. N. (2004). Toward a Model of the Effective Transfer of Scientific Knowledge from Academicians to Practitioners: Qualitative Evidence from the Commercialization of University Technologies. *Journal of Engineering and Technology Management, 21*(1–2), 115–142.

Singh, H. (2006). Prospects and Challenges for Harnessing Opportunities in Medicinal Plants Sector in India. *Law, Environment and Development Journal, 2*(2), 196, 198–210. http://www.lead-journal.org/content/06196.pdf.

Srivastava, V., Negi, A. S., Kumar, J. K., Gupta, M. M., & Khanuja, S. P. S. (2005). Plant-Based Anticancer Molecules: A Chemical and Biological Profile of Some Important Leads. *Bioorganic & Medicinal Chemistry, 13*, 5892–5908. Elsevier.

Taylor, G. (2010). Indian Sandalwood Plantations in Australia. In *Head of Plantations, Tropical Forestry Services Limited, Australian Forest Growers Conference 2010, Australia Forest Growers*, 107–109.

Teel, W. (1985). *Tree Seed Training and Extension Resources; A Pocket Directory of Trees and Seeds in Kenya* (151 pp.). Illustrated by Terry Hirst. Kengo, Nairobi. http://agroforesttrees.cisat.jmu.edu/.

The Guardian. (2004, September 5). The Multi-Billion Bio-Piracy Law Suit Over Faded Jeans and African Lakes (Antony Barnett, Public Affairs Editor, Sunday). https://www.theguardian.com/uk/2004/sep/05/highereducation.science.

UNESCO. (1998). *FIT/504-RAF48 Terminal Report: Promotion of Ethnobotany and the Sustainable Use of Plants Resources in Africa* (p. 60). Paris: UNECSO.

WHO. (2003). *Traditional Medicine* (Fact Sheet No. 134) (Revised May 2003). WHO: Switzerland. http://apps.who.int/gb/archive/pdf_files/WHA56/ea5618.pdf.

WHO. (2009). *Monograph on Selected Medicinal Plant* Vol. 4, (p. 1–456). Salerno Paestum, Italy: WHO Consultancy of Selected Medicinal Plants. ISBN 9789241547055.

CHAPTER 9

Conclusion: The Way Forward

Njoki Nathani Wane and Kimberly L. Todd

Abstract This chapter extends the discourse in this anthology by posit-
ing that the soul is essential to lasting decolonial change. It takes up the
question of how to transform the public sphere from the starting point
of the soul's role in healing colonial trauma. It seeks to achieve this task
by mapping the trajectory of the soul from a point of decolonial awakening to a full decolonial renewal of the spirit that has lasting implications
on the society. Decolonial futurities are invoked in this chapter, pointing our pathways towards communal revitalization of our planet. This
revitalization is based on love, spirituality and decolonial resurgence.
Liberation from colonial trauma and logics is at the heart of this chapter.
Concepts that are discussed include the Cartesian separations, ancestral
arc, re-membering and healing.

Keywords Ancestral arc · Healing · Colonial trauma · Soul ·
Liberation

N. N. Wane (✉) · K. L. Todd
Ontario Institute for Studies in Education, University of Toronto, Toronto,
ON, Canada
e-mail: njoki.wane@utoronto.ca

K. L. Todd
e-mail: k.todd@mail.utoronto.ca

© The Author(s) 2018
N. N. Wane and K. L. Todd (eds.), *Decolonial Pedagogy*,
https://doi.org/10.1007/978-3-030-01539-8_9

The complexities involved in this text were quite evident as authors moved in and out of different aspects of their chapters. Colonization is a phenomenon that has affected the whole world including colonized people and colonizers alike (Memmi 1965). We require modes of resistance, healing and reawakening in light of colonial traumas. Finding ways to honour the voices, cosmovisions and worldviews that make up the collective story of humanity, which is necessary for a decolonial transformation, is part of our vision. This book takes up the question of what is inherently oppressive and colonial about the places in which we reside, the places in which we learn and the places in which we heal and work? Out of these sites of colonial oppression, how can we continue to resist, resurge and renew our spirit, our minds, our communities and our world? This book explores these questions and makes a bid for liberation by engaging decolonial pedagogy in the examination of the public sphere.

The message in this book allows the reader to critically examine the challenging of state power through a spatial analysis of a Toronto neighbourhood and a nearby university, subsequently by exploring the psychological sciences' inability to engage decolonial frameworks despite colonial epistemic violence, by engaging the need for anti-colonial and decolonial education in school systems, by demonstrating the necessity of art education as medium of liberation and by unveiling the inherent wisdom and value in Indigenous technologies. This book has demonstrated a global mapping across disciplines and geographical locations of both hegemonic and colonial oppressions and sites of resistance, resurgence and renewal.

Reading each chapter was a reminder that our book is not a single story (Adichie 2009) but multiple stories that engages in rupturing colonization in order to engage in decolonizing the public sphere. Although each story had a unique message, it became clear that they were part of the collective. This in itself has forced us to start asking ourselves how do we now move from decolonizing pedagogies to ways of challenging the public sphere as a whole? What strategies would be required to engage in such work?

This is our reflection on the path of the soul moving through the phases of the colonial trauma and on a pathway towards liberation because we believe that the soul's journey is critical for bringing about lasting and sustainable decolonial change. Decolonization happens internally at the level of spirit and it ushers your mind, body and soul through a process that peels off the layers of colonialism. Colonization as a project seeps into an individual and becomes part of you without you realizing that it is a foreign entity within you. It is when the spirit comes in and shakes the foreign entity within the individual that the internalized decolonizing journey begins because there is the realization that an alien malignancy lives within you causing dis-ease and illness. This manifestation of colonial trauma can be psychological, spiritual and physical. So now the spiritual journey forged from ancestral and planetary past begins and resurges to transform the individual from the dormant state to an active state of awareness charting a pathway to the future. It is important to note that this pathway is unclear as it resides in a space of uncertainty and will continue to do so. You can only see a few feet ahead of you at any given time because what is required is faith, resonance and being attuned to the Mysteries of the Universe. A path lies ahead that requires deep courage, but it flows from a force within you and the force within you connects the inner self to the outer world, bridging transformation—we will refer to this bridging as the ancestral arc. The ancestral arc unleashes your ancestral origins (physical, cosmical, Earthly, familial and spiritual) and ignites the process of re-membering/remembering. We imagine it is a branch from a tree that starts at the brain stems and arcs up and above the soul seeker and shooting ahead to an unseen point in the future in which the person walks boldly towards. The branch is used as a way to symbolize a branch of the world/cosmic tree that connects us all past, present and future. Through the ancestral arc the rupturing of old and oppressive colonial systems and structures begins and pathways and opportunities arise. This decolonial work requires courage and fortitude because boundaries will be broken and new pathways charted.

This courage comes from the awakening of the soul to fallacy of the belief in the Cartesian separations (which dictate the separation of mind and body, mind and nature, matter and spirit) (Grosfoguel 2011). A vast number of western hegemonic structures are modelled off of this fallacy and that is why in this conclusion we are advocating for the awakening of the multiple dualities that exist in the universe because we the individuals are the universe. Yes, part of the universal energy connects all of us into spiritual being; through this communion and connection we are made whole.

The future involves our collective decolonial and spiritual transformation. This process requires the awakening of the collective consciousness by acknowledging the fragmentation and disconnection that colonialism continually enacts and perpetuates in our minds, in our homes and in our societies. Through this slow and painful realization something deeper calls to us, a calling that is both within and outside of ourselves. It may come in the form of dreams, visions, or synchronicities played out in the waking world and over time clarity will come. The ancestral arc moving us individually and collectively ever closer to our wholeness as one person's decolonial and spiritual transformation sparks another. Moreover, collective decolonial and spiritual transformation necessitates our commitment to work towards awakening the spirit and dispelling dis-ease in us. By dis-ease we refer to the physical, mental, emotional and spiritual alienation that originates from a lack of wholeness and belonging. Healing being an intrinsic component to our awakening. Part of this process lies in accepting the responsibility to continually value our calling by seeking guidance from all the realms—material and spiritual and their interface, since love both grounds and propels us. We need to re-connect to our planetary ancestry, for inherent within the Earth Herself lies our understanding of the value of diversity and the necessity for reciprocity, herein unfolds our origins and our pathway to collective decolonial futures.

There will be times when we will encounter derision, sabotage and derailment that are both internal and external. These moments of struggle provide the spring board for an energized, more focused and more situated path from both the spiritual and the Earthly realm. The place we return to in our moments of doubt, desperation and weaknesses is our Mother the Earth. She provides us with continual sustenance, support and love in an eternal embrace. For blue is the ancient colour of love (Meyer 2016), we are encased above and below in love as we stride into the decolonial future. We believe in the role that spirituality plays as a transformative mode of knowledge production. Spirituality as a transformative tool is a multidimensional, relational way of being in the world where one is connected to all that there is. The relational aspect of being *in* the world and *with* the world is eloquently articulated by Wilson (2008) who stated: "It is with the cosmos; it is with the animals, with the plants, with the earth that we share this knowledge. It goes beyond this idea of individual's knowledge..." (p. 74). What we are advocating for as we look into the horizon, is the importance of the *totality of being* in order to carry out this work. It is transcending the degradation of the human spirit that suffers under the yoke of continual busyness that plagues our everyday lives and being attuned and present to the Cosmos that is. This entails being able to rely on intuition and resonance, drawing from the messages encoded in our dreams, being able to read the landscape and remain open hearted and attuned to our spiritual and ancestral knowledges. Those who seek and yearn for a balance, who understand that we are in communion with ensouled dimensions of Earth and Cosmos are actively participating in the co-creative potential of the universe and will be able to stay true to the liberation course. The way forward is going to be challenging and barricaded by both visible and invisible—whether it be our colonial trauma, or our own internal critics or the systems and structures that surround us. We know we are supported in our work through our ancestry, by the Earth Herself, by the Cosmos and by staying true to our course. We are ready and prepared to join hands with whoever is willing to participate in this multifaceted liberatory praxis that we believe is attuned with decolonial work. Let us walk together towards a decolonial future of a truly thriving planet for all our relations, seen and unseen.

REFERENCES

Adichie, C. N. (2009, July/August). *The Danger of a Single Story*. Retrieved April 30, 2018, from https://www.ted.com/talks/chimamanda_adichie_the_danger_of_a_single_story.

Grosfoguel, R. (2011). Decolonizing Post-Colonial Studies and Paradigms of Political Economy: Transmodernity, Decolonial Thinking, and Global Coloniality. *Transmodernity: Journal of Peripheral Cultural Production of the Luso-Hispanic World, 1*(1).

Memmi, A. (1965). *The Colonizer and the Colonized*. Boston: Beacon Press.

Meyer, M. A. (2016, August 11). *Connecting Education and Environment: Mobilizing Sustainability in Education, Research, Policy, and Practice Keynote Address*. The Sustainability Education Research Institute (SERI). Retrieved April 30, 2018, from https://www.youtube.com/watch?v=DoGmHSwOpSA.

Wilson, S. (2008). *Research Is Ceremony: Indigenous Research Methods*. Winnipeg: Fernwood Publishing.

INDEX

© The Editor(s) (if applicable) and The Author(s) 2018
N. N. Wane and K. L. Todd (eds.), *Decolonial Pedagogy*,
https://doi.org/10.1007/978-3-030-01539-8

143

The manufacturer's authorised representative in the EU is Springer
Nature Customer Service Centre GmbH, Europaplatz 3, 69115 Heidelberg,
Germany. If you have any concerns regarding our products, please
contact ProductSafety@springernature.com

Printed and bound by CPI Group (UK) Ltd, Croydon, CR0 4YY
29/04/2026
02099478-0017